A Chemist's Secret to Cake Baking

Walter H. Chan Ph.D.

TABLE of CONTENTS

ACKNOWLEDGEMENTS

I want to thank Ms. Kirsten Mania and Ms. Esther Yu for their critical review of the manuscript and their comments.

I am grateful to Ms. Valerie Lam for her generosity in sharing her expertise in communications and photography. Many of her valuable insights have appeared in the book's format and text. Her photographs have added clarity to the baking equipment discussion and great appeal to the cake recipes.

Finally, I am indebted to Mr. Derek Christopher for volunteering his graphic design talents in creating the baking framework figure, the baking equipment collage, and the book cover. His typesetting has also greatly enhanced the look and feel of the book.

Without the contribution of all these individuals, the completion of this book would not have been possible.

PREFACE

This is not just another cake recipe book.

In year 2000, I picked up baking as a hobby while in fulltime work and in fulltime study. To me, it was therapeutic and enjoyable. As a novice baker then, I also had my frustrations, being unable to produce wonderful baked goods anywhere resembling the recipe pictures. Partly it was because, in many recipes, ingredients were not adequately standardized and steps not quantified and partly it was because baking equipment varied.

I believe that my experience is not unique. Bakers desire consistent and reproducible results. Understanding the "what's" and the "why's" allows bakers to make meaningful variations in the "how's". Thus, I was determined to remove the uncertainties often encountered in recipes available to us by marrying baking and science. Using my training as a scientist, I have prepared this book.

The book is unique in that it uses a chemist's analytical approach to understanding baking, that is, taking a rational approach by asking questions and doing experiments in baking.

There are three parts in this book.

Part I   is about the fundamentals of baking. I have introduced a baking framework describing how baking ingredients, baking process, and baking equipment come together contributing to the quality of baked goods.

Part II   is a quick reference source for bakers. It includes at-a-glance summary tables, baking measurement conversions, frequently encountered baking problems, and a checklist for successful baking.

Part III   is a sampling of a number of proven recipes. The 12 selected recipes are recipes that I have personally developed and tested. Noteworthy is that I have standardized the ingredient measurements and quantified preparation times.

Depending on your baking experience, you may choose to read the book sequentially or to proceed to cake recipes in Part III and then read Part I and Part II later. You will also get a lot of useful information by having a quick read of the baking framework figure in Part I and summary tables in Part II, leaving the text for future resource.

Baking is an art as well as a science. If you aspire to quality baking, this book is for you. By providing you with the "what's" and "why's" in this book, I hope you will succeed in the "how's".

Enjoy your new baking journey and be proud of your baked products!

PART I:
UNDERSTANDING YOUR CAKE BAKING LIKE A CHEMIST

A baker's goal in baking is to produce quality baked goods in both structure and texture. Structure is how ingredients within baked goods are organized and texture is the sensation cakes impart when consumed. Quality cakes appeal to the eyes, to the nose, and to the mouth. To the eyes, it includes colour, shape, and volume. To the nose, it is aroma. To the mouth, it is taste and mouth feel.

In order to succeed in baking, every baking step is crucial, be it the selection of ingredients, the execution of preparation steps, or the choice of equipment.

There are three chapters in this segment of the book and they cover baking ingredients, baking process, and baking equipment. Together they provide a good foundation to the basics of baking, the "whats" and the "whys", that form the basis to the "hows". To increase a baker's chance of success and to create a strong sense of accomplishment, the use of and substitutions to baking recipes must be done through a solid understanding instead of a hit and miss approach.

I BAKING FRAMEWORK

Flour is the most basic ingredient used in baking, giving baked goods their structure and texture. Most baking ingredients fall under two broad categories; those that toughen (strengthen) baked goods and those that tenderize (weaken) baked goods. It is the balanced use of the two types of ingredients that yields the desired structure and texture.

There is yet a third category; other ingredients that neither toughen nor tenderize baked goods but they contribute to the overall enjoyment.

A baking framework has been introduced here (Figure 1) to help illustrate this ingredient concept.

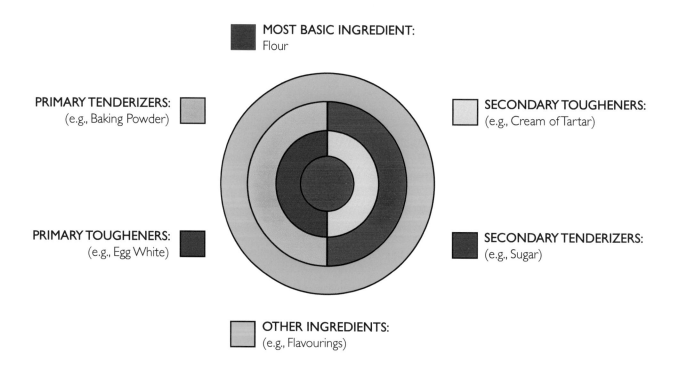

Figure I: A Baking Framework

In addition to the choice of baking ingredients, i.e., subject of this chapter, successful baking depends also on the appropriate execution of baking process and the proper use of baking equipment. They are the subjects of Chapters 2 and 3 respectively.

BASIC INGREDIENT
FLOUR IS THE MOST BASIC INGREDIENT IN RECIPES OF FLOUR CAKES.

Flour

Flour toughens and provides structure to baked products. Flour contains proteins and carbohydrates (starch) primarily. Glutenin and gliadin (two of over 30 kinds of protein in flour) form gluten upon contact with liquid(s), such as water. Glutens are elastic and stringy strains that trap and hold air bubbles. The starch in flour absorbs water, gelatinizes, and sets as it heats, adding to and stabilizing baked goods' structure. Air bubbles in the batter expand upon heating through vapourization of liquid(s). When the bubbles come in contact with carbon dioxide gas released from a chemical leavener, they will be expanded, causing the batter to rise and giving structure to baked goods.

Too much flour and/or mixing give rise to too much gluten, resulting in tough cake texture. Too little flour and/or mixing do not produce enough gluten, resulting in collapsed cakes. The overall quality, that is, structure and texture, of baked goods reflects the balanced use of other baking ingredients (than flour) that strengthen or weaken the structure.

Various types of flour are used for different purposes. They contain varying amounts of proteins that affect the elasticity or rigidity of baked products. Flour type reflects the kind of wheat from which the flour is milled. High gluten flours are from hard, winter wheat and low gluten flours are from soft, spring wheat. Bleached flour, having a lighter colour, yields more tender baked goods while unbleached flour, having a darker colour, yields more crisp baked products. Bleached flour and unbleached flour, however, have the same nutritional values.

Flour stored in an airtight container in cool temperatures should be good for six-months or longer. Some recipes call for the use of different combinations of all purpose and cake flours. Types of flour commonly used in baking are:

ALL PURPOSE FLOUR: having 11.0 – 11.8 per cent protein, is milled from endosperm (inner portion) of wheat grains after the bran (outer coat) and the germ (seed embryo) have been removed. It is milled from hard wheat or a blend of hard and soft wheats.

ALMOND FLOUR: some cake recipes call for almond flour as a replacement for wheat flour. Almond flour that has a consistency more like corn meal than wheat flour is made from ground blanched (i.e., no skin) sweet almonds.

BREAD FLOUR: containing high protein (11.7 – 13.2 per cent) and low starch contents, is milled from blends of hard red wheats and is specifically formulated for making yeast bread.

CAKE FLOUR: containing 6.7 – 8.0 per cent protein, is milled from soft white wheat and is suitable for cakes of which a delicate, tender texture is desired. Almost all cake flours are bleached.

CAKE MIX: some cake recipes use cake mixes instead of flour as the basic ingredient. The use of cake mixes simplifies preparation and increases the chance of success for the novice. These recipes add other ingredients and introduce variations to the cake mixes' preparation steps. This results in quite delicious cakes and the use of cake mixes is not easily recognized. Chocolate and yellow cake mixes are popular choices.

PASTRY FLOUR: with 9.0 – 10.6 per cent protein, is made from soft wheat and is used for pastry baking. It has less starch than cake flour.

In the United States, cake flour and pastry flour can be purchased as separate entities. In Canada, however, it is the combined "cake/pastry" flour that is sold in supermarkets.

SELF RISING FLOUR: is flour mixed with baking powder and salt in the proportion of 1 c flour : 1½ to 1¼ tsp baking powder : a pinch to ½ tsp salt. On the one hand, it is for convenience; but on the other hand, it has the drawback of deteriorating quickly upon exposure to humid conditions.

WHOLE WHEAT FLOUR: consisting of 12.5 – 13.5 per cent protein, is brownish, coarse textured flour prepared by grinding the whole wheat kernel, including bran, germ, and endosperm. It is more nutritious and has more fibres than all purpose flour. Baked goods made from whole wheat flour are denser and heavier compared to those made from white flour. The presence of bran in wheat flour reduces gluten formation. In addition to carbohydrates and proteins, whole wheat flour also contains a little bit of fat.

TOUGHENERS
TOUGHENERS (ALSO REFERRED TO AS STRENGTHENERS) CAN BE GROUPED INTO PRIMARY TOUGHENERS AND SECONDARY TOUGHENERS.

PRIMARY TOUGHENERS
PRIMARY TOUGHENERS ARE THOSE INGREDIENTS THAT HAVE THE TOUGHENING FUNCTION WITHOUT THE HELP OF OTHER INGREDIENTS.

Egg (whole)

Whole eggs are one of the most used ingredients in baking. For fat-containing (shortened) cakes, whole eggs are used to strengthen baked goods' structure because they consist of proteins. Beating whole eggs result in trapping of air bubbles that, upon baking, expand and leaven baked goods. The extent of air bubbles from beating whole eggs is much less than from beating egg whites alone. Eggs are nutritious; they contain proteins, fats, minerals, and vitamins. They also bind other baking ingredients together as well as add flavour and colour to baked goods.

Many recipes use eggs without specifying their size explicitly. It is reasonable to assume that they are large eggs. Preferrably, the size of egg should be specified in recipes to avoid guessing.

Egg White

Beating egg whites alone is a crucial step for baking un-shortened (foam) cakes, such as angel food cake and sponge cake. When egg whites are beaten, albumen (the protein in them) is stretched, forming a foamy, elastic film network that traps air bubbles. During baking, the egg protein denatures and coagulates around the air bubbles, giving structure to the foam. Air is released, causing the batter to rise and producing a light and airy texture of baked goods.

Egg whites of fresh eggs are tight and there is a transparent jelly around the yolks. As eggs age, egg whites become more fluid, which facilitates air incorporation, producing more volume upon whipping, a welcome feature for foam cakes.

SECONDARY TOUGHENERS
SECONDARY TOUGHENERS ARE THOSE INGREDIENTS THAT FACILITATE THE TOUGHENING FUNCTION OF OTHER INGREDIENTS.

Cream of Tartar

Cream of tartar makes egg proteins coagulate faster and helps maintain beaten egg whites' fluffiness and full volume. It is also a component of baking powder. When cream of tartar comes in contact with liquid(s), it reacts with baking soda to release carbon dioxide, causing baked goods to rise. Cream of tartar is made from grapes and is a byproduct of wine making.

Fat and Sugar

When fat, e.g., butter, is creamed with sugar, it incorporates air cells, producing good volume and fine and airy texture. When over-beaten, it loses its ability to trap air. Under-creaming, however, results in coarse grain and lumpiness of baked goods.

Liquid

Liquids may be used as a toughener or as a tenderizer. See also "Liquid" under "Primary Tenderizers" and "Secondary Tenderizers".

WATER AND OTHER LIQUIDS: liquids, such as water, hydrate flour's proteins to form gluten and starch to gelatinize, bringing about baked goods' texture and structure. They dissolve other baking ingredients, such as sugar and salt. Water can also aid to increase the volume of egg white foam; however, too much water decreases its stability. Water is a possible partial substitute for oil for those who are calories conscious.

Salt

Salt strengthens glutens, producing good grain and texture for baked products. In addition, salt helps stabilize beaten egg whites' volume; however, some bakers claim that adding too much salt may decrease beaten egg white foam's stability. A small amount of salt (say, ⅛ tsp) in the recipe also enhances the sweetness of sugar in the batter and adds flavour to the other ingredients of baked goods without its saltiness detected.

Sugar

Sugar interacts with egg proteins to stabilize whipped egg white foam structure to create the right texture and volume. Furthermore, sugar also adds flavour and aroma, causes browning, and improves keeping qualities of baked goods. Browning occurs both as a result of caramelization (i.e., thermal decomposition of sucrose into fructose and glucose) and Maillard reaction (i.e., a reaction involving the carbon molecules in the sugars with the amino acids of the proteins, with the addition of heat). See also "Sugar" under "Secondary Tenderizers".

Sugar is extracted/purified from sugar canes and sugar beets. Sugar exists in many forms.

ARTIFICIAL SUGAR: is a non-caloric sweetener. It is typically 200-300 times sweeter than sucrose that is found in granulated white sugar.

BROWN SUGAR: is made from refined white sugar and molasses (sugar syrup). Light brown sugar and dark brown sugar contain 3.5 per cent and 6.5 per cent of molasses respectively.

CONFECTIONER'S SUGAR: is used primarily for icing and glazing cakes. It is sugar that has been ground into a fine powdered sugar with about 3 per cent starch mixed into it to prevent it from clumping. It is also called icing sugar.

GRANULATED SUGAR: is processed granular white sugar (sucrose). It is the most commonly used sugar in baking.

HONEY: is used in some cake recipes instead of sugar. Honey, a liquid sugar made by bees, is sweeter and has more vitamins and minerals than sugar. It not only raises a person's blood sugar slower than sugar, but also keeps baked goods moister longer. In substituting sugar by honey in recipes:

o   use 1:1 ratio for up to one cup of sugar; however, use no more than ¾ : 1 when more than a cup of sugar is called for;

o   reduce 25 per cent of liquid called for in the recipe;

o   add ½ tsp of baking soda (to neutralize honey's acidity) for every cup of honey used;

o   lower baking temperature by 25°F to control browning.

SUPERFINE SUGAR: its use is desirable to achieve a finer texture of baked goods because it is readily soluble in liquids. As well, when creamed with fats (e.g., butter), it allows more air to be trapped because of its finer size and more contact area. Superfine sugar (called castor or caster sugar in Great Britain) can be prepared from grinding granulated sugar with a coffee bean grinder or a food processor.

TENDERIZERS
TENDERIZERS (ALSO REFERRED TO AS WEAKENERS) CAN BE GROUPED INTO PRIMARY TENDERIZERS AND SECONDARY TENDERIZERS.

PRIMARY TENDERIZERS

PRIMARY TENDERIZERS ARE THOSE INGREDIENTS THAT HAVE THE TENDERIZING FUNCTION WITHOUT THE HELP OF OTHER INGREDIENTS.

Dairy Product
Dairy products are often added to cake batter to increase moistness of baked goods. Many of them are slightly acidic and react with baking soda to form carbon dioxide, which functions to expand air bubbles that exist already in the batter (see "Leavener" later).

CREAM CHEESE: made from a combination of cheese and milk, is a soft, sweet, mild-tasting white cheese. It is used in making cheese cakes, tiramisu, and carrot cake icing. In Canadian supermarkets, cream cheese products are available in different fat contents, ranging from low fat (95% fat free), light fat (14% M.F.) to regular fat (up to 31% M.F.) Low fat cream cheese has much less fat contents but is not as creamy as their light and regular counterparts.

MILK: is available in various fat contents, i.e., whole (or homogenized), 2 per cent, or skim. Milk contains proteins, fats (except skim milk), and other nutrients. Not only does milk add flavour, it (when containing fat) also helps browning of baked goods. Milk is a possible substitute, to some extent, for oil for those who are calories conscious.

Many recipes use milk without specifying its type explicitly. It is reasonable to assume that whole milk or 2 per cent milk is called for. Preferrably, the type of milk in recipes should be specified to avoid guessing.

BUTTERMILK: is the sour-tasting liquid left over after milk or cream has been churned to make butter. Evaporated milk (with 60 per cent water removed), condensed milk (evaporated milk with large amounts of sugar, i.e., 40-45 per cent, added), and coconut milk are also used in some recipes such as cassava cakes.

SOUR CREAM: is a thick smooth cream that has been soured artificially. Sour creams are available commercially in different fat contents and, in Canadian supermarkets, typically vary from 1 per cent to 14 per cent. Sour cream can usually be substituted by yogurt in cake baking.

YOGURT: is made from bacteria-fermented milk and it has a thick and creamy consistency. Yogurt is available in various fat contents in Canadian supermarkets; typically varying from zero to 3 per cent fat. While plain yogurt is often used in baking, the use of flavoured yogurt, such as that with a vanilla flavour, creates a surprisingly unique taste in baked products. Yogurt can usually be substituted by sour cream in cake baking.

Egg Yolk

Egg yolk emulsifies fats and liquids due to its lecithin content. It also contributes to the flavour, colour, and texture of baked goods. When egg yolks are beaten, they also trap air bubbles but to a much lesser extent as egg whites.

Fat

Fats are often used in baking recipes. They enhance moistness of baked goods, making the crust tender. Some recipes call for various combinations of shortening and butter (or margarine). For foam cake recipes such as sponge cakes, fats (e.g., butter and margarine) should not be used.

Fat coats the proteins in flour and decreases their contact with liquid(s), hence reducing gluten formation and over-structure of baked products. The fat's coating makes gluten strands slippery, allowing gas bubbles to move easily and resulting in baked goods' finer texture. Fats are also called shorteners because they shorten gluten strands. Fats, such as butter and margarine, add flavour to cakes.

BUTTER: is a dairy product made by churning fresh or fermented milk or cream. It is used in many recipes to add a sweet, delicate, and rich flavour to baked products that margarine, vegetable oil, and other shortenings do not offer. Unsalted butter rather than salted butter should be used in baking since the latter includes an amount of salt that cannot be easily quantified. For better results, shaved butter left standing in room temperature is preferred to liquid butter, e.g., melted by heat, for mixing or creaming.

Many recipes use butter without explicitly specifying its type. It is reasonable to assume that unsalted butter is called for. Preferrably, the type should be specified in recipes to avoid guessing.

MARGARINE: is sometimes used as an alternative to butter. It is partially hydrogenated fat or oil (of animal or vegetable origin), together with water, milk solids, and salt. Margarine has less saturated fat. It does not have the same flavour and aroma that butter offers. Some bakers prefer stick margarine to spreads and whipped margarine since the latter contain more air.

SHORTENING: is a solid form of vegetable fat. Shortening is white and does not add flavour to baked products. Shortening has an advantage over other fats since it contains millions of nitrogen bubbles infused during production and holds bubbles well. These bubbles help enhance the leavening and lightening of the batter, producing softer, thicker, and chewier baked goods.

Leavener

A leavener is a substance that causes batter to rise through expanding the air bubbles already present in it. This gives baked goods a light and porous texture and tender crumb.

Three leavening processes give rise to leavening gases: biological (yeasts to generate carbon dioxide; mainly used in bread making), heating (liquids, such as water, form vapour during baking), and chemical (mainly used in cake baking to produce carbon dioxide). The first two types of leavener will be discussed only briefly here.

When fermented, moistened, or heated, carbon dioxide or steam is formed. This process expands the air bubbles already exist in the batter through gluten formation, whipped egg whites, or creamed sugar and fat. Adding too much leavener creates supersized bubbles that burst, leading to a coarse, fragile crumb and a fallen centre. Adding too little leavener does not generate enough leavening gas, resulting in a tough cake with poor volume and compact crumb. More leavener can be used, however, when baking recipes call for larger quantities of heavy ingredients, such as chopped dried fruit and nuts.

Sift and mix powdered leavener (i.e., baking powder or baking soda) with flour before adding them to the batter to ensure uniformity and to avoid large holes in baked products. Keep leavener under cool temperatures in an air tight container.

BAKING POWDER: is a chemical leavener containing baking soda ($NaHCO_3$ - a base; see below) and one or more dry acids. It is mixed in cornstarch to keep it dry and running.

When baking powder comes in contact with liquid(s), it triggers an acid-base reaction, leading to the release of carbon dioxide and causing the batter to rise.

$$NaHCO_3 + H^+ \rightarrow Na^+ + CO_2 + H_2O$$

BASE   ACID ION   SODIUM ION   CARBON DIOXIDE   WATER

Depending on the acid(s) that it contains, baking powder may be a single-acting or a double-acting type.

For single-acting baking powder, baking soda reacts with an acid that is fast dissolving at room temperature. Cream of tartar and monocalcium phosphate are examples and the acid-activated reaction takes place immediately upon contact with liquid(s). Therefore, when using single-acting baking powder, the batter must be baked as soon as possible.

Double-acting baking powder contains other acids that dissolve slowly, e.g., sodium aluminum pyrophosphate, or dissolve only at higher temperatures, e.g., sodium aluminum phosphate. The release of carbon dioxide therefore takes longer.

In addition to the above chemical reaction, sodium bicarbonate breaks down also at temperatures above 120 °C (250 °F), without any acids involved, releasing carbon dioxide.

$$2\, NaHCO_3 + heat \rightarrow Na_2CO_3 + CO_2 + H_2O$$

SODIUM BICARBONATE   SODIUM CARBONATE

Only double-acting baking powder is available in today's commercial market. Single-acting type baking powder is available only to manufacturers.

Watch for the expiry date for efficacy. If in doubt, baking powder should be tested by mixing ½ tsp of it with ½ c of just boiling water. If it does not bubble fiercely, the baking powder should be discarded.

BAKING SODA: is sodium bicarbonate ($NaHCO3$), a base. When it comes in contact with a liquid that is acidic (such as buttermilk, fruit juice), an acid-base reaction takes place, forming carbon dioxide gas (CO2). It is the release of CO2 that causes baked goods to rise.

Using baking soda alone in cake recipes without baking powder is not common. By itself, baking soda does not have leavening capabilities. Baking soda is used in recipes containing an acidic ingredient such as buttermilk, yogurt, and sour cream. It functions to neutralize the acids in the recipe and to enhance leavening, adding tenderness to baked goods. For recipes containing only baking soda without baking powder, the batter should be baked as soon as possible before its leavening power is diminished.

Baking soda is also used in chocolate cakes to raise the pH of cocoa to help acquire a reddening colour of the cake, e.g., Devil's food cake. Too much baking soda not only damages the volume and texture of a cake, making it coarse, but also results in a soapy, bitter taste. Testing of baking soda can be done by adding ½ tsp of it to ½ c of vinegar. If it does not fizzle, it should be discarded.

LIQUID: Liquids cause batter to rise upon vapourization during baking, thus expanding the air cells already exist in the batter. See also "Liquid" discussions under "Secondary Tougheners" and "Secondary Tenderizers".

YEAST: is a biological agent (one-celled plant) that is used primarily in bread making. Optimum yeast development (i.e., to ensure the yeast is active) occurs under appropriate proofing conditions, including a warm temperature (110 – 115 °F or 43 to 46 °C) and a high humidity. Fermentation releases carbon dioxide and alcohol, affecting the volume, flavour, and aroma of baked goods.

Liquid
Liquids provide moistness for baked goods. Be careful; too much liquid causes baked goods to become too heavy and collapse. (See also discussions of "Liquid" under "Secondary Tougheners" and "Secondary Tenderizers").

JUICE: such as orange juice and lemon juice, is often used in cake recipes. For some recipes, if appropriate, juices may substitute water to bring out the fruit flavour.

VEGETABLE OIL: oils coat flour particles, reducing gluten formation. Vegetable oils are not fats and are used in baking foam cakes such as sponge cakes. They don't trap air bubbles when beaten with sugar as fats (e.g., butter) do. When used in shortened cakes, oils do not make fine grained baked goods and their texture is heavier.

WATER: water is sometimes used in recipes (especially when cake mixes are used as the basic ingredient) to provide a leavening function.

Pudding Mix
Some cake recipes, in addition to using only flour or cake mixes, use also pudding mixes. Pudding mixes, upon contact with liquid(s), provide moistness to baked goods. Common flavours are banana, chocolate, and vanilla.

SECONDARY TENDERIZERS

SECONDARY TENDERIZERS ARE THOSE INGREDIENTS THAT FACILITATE THE TENDERIZING FUNCTION OF OTHER INGREDIENTS.

Dairy Product

BUTTER MILK, SOUR CREAM, AND YOGURT: they are used in some recipes to act as an acid to react with baking soda (a base) to form carbon dioxide (CO_2), thus serving also a secondary tenderizer function.

Liquid

LIQUID: liquids that are acidic, e.g., fruit juice, react with baking soda (a base) to form and release carbon dioxide, expanding already trapped air bubbles in the batter. This causes the batter to rise, resulting in light and porous texture, good volume, and tender crumb.

Sugar

In addition to having a secondary toughening function (See discussion of "Sugar" under "Secondary Tougheners"), sugar also has a secondary tenderizing function in baking. It provides tenderness for baked goods by competing for liquids required for flour to form gluten to trap air bubbles and by absorbing water required for starch to gelatinize to set structure.

OTHER INGREDIENTS

THESE ARE INGREDIENTS THAT ARE NEITHER TOUGHENERS NOR TENDERIZERS.

Flavouring

CHOCOLATE: is a favourite ingredient in many cake recipes. Chocolate is produced from cocoa (or cacao) beans. For baking purposes, unsweetened chocolate, which is solid chocolate liquor, is most commonly used. Semi-sweet chocolate (made from chocolate liquor, cocoa butter, sugar, and vanilla flavouring) and white chocolate (having no chocolate liquor but containing cocoa butter, milk solids, milk fat, and sweeteners) are also favourites. Chocolates can be purchased in cubes, bars, and buds (chips).

CINNAMON: ground cinnamon is used in some recipes such as carrot cakes to add its aromatic and sweet flavour.

COCOA POWDER: is produced from the low-fat component of chocolate. Cocoa powder is acidic and has a sour taste. It is more reddish than chocolate unless treated by Dutch process to neutralize its acidity, giving it a brown chocolate colour and a milder taste. Other than its subtle flavour, cocoa powder is also used to give chocolate cakes their chocolate colour and to coat baking pans for chocolate cakes.

EXTRACT: many recipes call for the use of almond, lemon, orange, and vanilla extracts, etc. Extracts are extracted oils mixed with water in alcohol. Both natural and artificial extracts are available commercially. Natural extracts are from plants or seeds. Vanilla extract is most commonly used. A combination of different types of extracts brings about unique flavours in cakes such as chiffon and sponge cakes; it is limited only by one's imagination.

LIQUEUR: is often used in recipes, such as Tiramisu, to add its unique flavour to cakes. Favourite liqueurs are Irish Cream and Tia Maria.

NUTMEG: is used in some baking recipes in its grated form, especially for Christmas baking, to add intensity and flavour to baked goods. Nutmeg is a spice produced from nutmeg tree's brown and wrinkly seeds.

Fruit

Fruit are often used to enhance the flavour of baked goods. Fresh and frozen fruit add moisture and dried fruit compete for moisture in the baking process. The use of large amounts of fruit may hinder the expansion of air bubbles. It is important to use small-sized pieces to avoid blockage of the batter from rising, resulting in flat and tough baked products. Fruit types are not readily interchangeable in recipes because of their varying moisture contents.

DRIED FRUIT: often offer aromas sometimes not available from fresh fruit. Too much dried fruit competes for moisture and too large-sized dried fruit blocks the batter from rising. Dried fruit such as apple, apricot, cranberry, and mango are often used in cake recipes.

FRESH FRUIT: are often used between cake layers and as cake toppings (for decoration) to add unique taste to baked goods. Sometimes they are also used as a key ingredient in recipes, e.g., banana cake. Blueberry, strawberry, kiwi, and mango are frequently used in cheesecakes and other cakes. Upon baking, sometime fresh fruit, such as mango, may lose their unique flavour, therefore pre-conditioning is necessary.

FROZEN FRUIT: are sometimes used as a substitute for their dry and fresh fruit counterparts. Defrost and drain excess fluids before use. Other than being more liquidly, some find frozen bananas sweeter; however, they retain their wonderful aroma when thawed. Some other frozen fruit, e.g., strawberry, however, do not taste the same compared to when they are fresh.

Fruit Zest

Some recipes (for instance, chiffon cakes) call for zests of fruit such as orange and lemon. Fruit zests accentuate fruit flavours from fruit extracts and provide a unique mouth feel. Prepare fruit zests from washed and dried fresh fruit. Take precaution not to grate into the pith or the fruit flavour will be compromised.

Nut

Nuts introduce a unique mouth feel to baked goods. In the baking process, they may block the expansion of air bubbles if their sizes are too large. It is important to use thin slices or small-sized pieces to avoid blockage of the batter from rising, resulting in flat and tough baked goods. Many baking recipes use nuts such as almond, hazelnut, pecan, and walnut. These nuts can be purchased in whole, half, or sliced forms. When kernels are moldy, shrivelled, or dry, discard nuts as they are usually rancid or bitter.

2 BAKING PROCESS

In addition to choosing the best possible ingredients, executing the baking process properly increases the chance of baking success. From the choice of ingredients to the finished cakes, there are a number of key steps, namely:

- ingredient preparation;
- batter preparation;
- batter transfer;
- baking;
- post-baking activities.

Each of the process steps is important and this segment of the book is devoted to them.

INGREDIENT PREPARATION

Egg Separation and Beating/Whipping

Separating egg white from egg yolk is necessary for foam (un-shortened) cake baking. Since traces of fat from egg yolks reduce whipped egg white volume, it is critical to have a clean separation. For better results, separate egg whites from egg yolks while eggs are still cold from the refrigerator. Proceed with preparation steps only when egg whites and egg yolks have reached room temperature after at least 30 minutes.

For egg yolks, beat on medium speed for two minutes with an electric mixer. Add sugar (superfine sugar preferred) in a steady stream along while beating for another three minutes on medium high speed. This will stabilize whipped egg yolks. Stop beating when yolks are thick and pale in colour, forming ribbons when beaters are lifted. Egg yolks will start to deflate after standing for five to eight minutes.

For egg whites, beat until fluffy. This usually means starting on low speed for one minute, on medium low speed for two minutes, and followed by four minutes on high speed. Over-beaten egg whites are too stiff and have no more stretch left to rise when baked. Under-beaten egg whites do not hold enough air and baked goods are flat. It is important to use low and medium low speeds at the early stages to avoid deflating the foam. Beaten egg whites start to deflate after five minutes and should be baked as soon as possible. Add a small amount of sugar and salt in (foamy) stage 1, and cream of tartar in (stiff peak) stage 3 to stabilize beaten egg whites.

Fruit and Nut Preparation

Common to dried, fresh, and frozen fruit and nuts is a pre-treatment before folding them into a batter: coating fruit or nuts with a small quantity of flour by shaking them in a zipped freezer/sandwich bag. It not only helps fruit/nut pieces separate better from one another, but also prevents them from sinking to the bottom of the batter. In addition, this thin coating of flour also absorbs some of the liquid that may be released during baking.

DRIED FRUIT: in order to have better moisture and hardness control, pour boiling water over chopped dried fruit (e.g., apricot, cranberry, etc.) and allow them to drain for five minutes. This helps ensure that dried fruit are tender and ready for folding into the batter.

FRESH FRUIT: some fresh fruit lose their aroma upon baking. In order to preserve their aroma, cut fresh fruit into small pieces and chill them by surrounding them with ice prior to folding them into the prepared batter.

FROZEN FRUIT: while usually not the preferred ingredient, frozen fruit are sometimes used as a substitute when needed. Defrost frozen fruit, cut them into small pieces, and rest them at room temperature before use. Use a paper towel to dry off excess moisture before folding fruit into the prepared batter.

NUTS: thinly slice or cut up nuts into small pieces.

Pre-Measuring Ingredients
Read and understand recipe requirements upfront and measure out ingredients prior to batter preparation. For example, measuring out and sifting flour before mixing, measuring out and cutting fruit before folding, etc. To ensure that you have all ingredients ready before preparation steps helps avoid missing ingredients and saves time.

Utensil Cleaning

Even with great recipes, small traces of contaminant can prevent you from success, e.g., trace amounts of fat can deflate the air bubbles in whipped egg whites for foam cake baking. It warrants taking special care in washing, rinsing, and drying baking equipment. Choosing the right utensil material is equally important. Plastic and wooden surfaces tend to absorb and retain greases and they are difficult to wash off, hence you should avoid using them for baking foam cakes.

BATTER PREPARATION

Ingredient Folding
When folding solid ingredients such as dried fruit, nuts, and chocolate chips into prepared batter, use a spatula in such a way that the batter will not be over-mixed. If a mixer is used in the process, be sure to use low speed.

A critical step in foam cake baking is folding beaten egg whites into prepared egg yolk/flour batter. Some bakers advocate folding a lower density component (e.g., whipped egg whites) into a higher density component (e.g., egg yolk batter). However, as long as you do it gently and steadily with a spatula, this is not a major concern and either method works well.

Ingredients Mixing/Creaming

Both over- and under-mixing of batter is undesirable. Over-mixing of flour forms too much gluten and breaks down batters, leading to deflation of baked products and making them tough. Under-mixing of flour does not create a sufficient gluten network to trap air and risks uneven mixing, leading to a weak structure and formation of lumps. Similarly, over-creaming butter and sugar leads to the loss of their ability to trap air and under-creaming results in lumpiness.

BATTER TRANSFER

Avoiding Overflow
To avoid batter overflow from baking pan into oven during baking, fill pan with batter to no more than 70 per cent full.

Choosing Baking Pan
Both baking time and baking temperature depend on baking pan's dimensions, shape, and material. It is important to use the baking pan specified in the recipe. However, if the specified pan is unavailable and substitution is necessary, refer to Tables 9 and 10 in Part II for guidance.

Greasing Baking Pan
Use a brush or paper towel to grease baking pans with butter, margarine, or oil. As an alternative, use vegetable oil spray. Be sure not to over spray because the excess oil around the batter may cause cakes to bake unevenly or burn. For better results, coat greased pan with a thin layer of flour (or cocoa powder for chocolate cakes). Alternatively, use parchment paper to line the bottom and sides of pan. Some spray the pan before putting on parchment paper to keep it in place and spray again on top of it for easy removal. Wax paper does not work as well with oil spray and that's why many prefer parchment paper to it.

It should be noted that pans for foam cakes should not be greased, as fats deflate beaten egg white foam. Also, lubricants, e.g., vegetable oil, make surfaces too slippery for foam cakes to cling to when rising.

Pre-conditioning Silicone Baking Pan
Cakes baked in silicon baking pans can be unmolded easily; however, uneven heating sometimes occurs in certain areas of the pan, causing batter on the sides of the pan to bake faster and possibly burn. The situation can be remedied by preheating the silicone pan in an oven for three minutes before pouring batter into it.

Transferring Batter
Gently transfer batter from a mixing bowl to a baking pan. To avoid tunnelling and to ensure even baking, shake or tap the batter-filled pan against a counter top or use a spatula to cut through the batter to help release large air bubbles.

BAKING

Choosing Baking Temperature and Baking Time

Under normal circumstances, stick to the recipe's recommended baking time and baking temperature. When modifying recipes, bear in mind this principle: choose baking time and baking temperature to reflect the amount of flour, pan dimensions, and pan material. In general, 325°F and 45-60 minutes (depending on pan type) are appropriate for recipes with up to 1½ c of flour. For recipes with more than 1½ c and up to 3 c of flour, increase baking temperature to 350°F. And for recipes with more than 3 c of flours, consider further increasing baking temperature to 375°F. Rarely baking temperatures above 375°F are desirable because of the risk of setting the outside of the cake too quickly but the inside is not baked properly. If large amounts of other dense ingredients are used, e.g., dried fruit and nuts, consider increasing baking time and/or baking temperature. For guidance on pan dimensions, pan type, and pan material, see Tables 9 and 10 in Part II.

Pre-heating Oven

Pre-heat oven for 15 minutes before baking to ensure the inside is evenly heated. Use an oven thermometer, e.g., a digital type, to confirm that temperature is even in the oven. Opening oven door during baking results in uneven temperature. Avoid doing it, especially in the early stages of baking.

Rack Positioning

For most cakes, use middle rack position for baking unless it is otherwise specified in recipes. In order to reduce uneven heating due to potential hot spots, place cake pans in the centre position of the rack.

Testing for Doneness

Test for a cake's doneness by inserting a wooden BBQ skewer stick or a piece of raw spaghetti into the cake and ensuring that it comes out clean. Test for doneness about three minutes before the recipe's specified time.

POST-BAKING ACTIVITIES

Cake Freezing

You can store most cakes in a freezer for future consumption. Wrap individual sliced cakes with a plastic sheet and keep them in a zippered freezer bag for storage in a freezer. Some cakes may retain their freshness and moistness even over a three-month period; however, storage efficacy varies from one kind of cake to another.

Cake Slicing

Nicely cut cake slices make baked goods presentable and appealing. Never use a blunt knife to slice a cake; it can destroy all the efforts gone into baking it. An electric knife does magic in that it cuts perfectly because it can cut fast through the cake. This is especially important with foam cakes because of their spongy texture. A second choice is a bread knife. Run the cutting knife under very hot water and dry it before use; it improves the quality of the cut. For those who like larger-sized cake slices, cut slantingly.

Cake Unmolding

For shortened cakes, after baking, leave cake pan on a wire rack to cool for 15 minutes before removing the cake from it. After proper cooling, run a thin, dull knife around the cake inside of the pan to loosen it from the sides of the pan. Peel off parchment paper, if used, from the bottom and sides of the cake. Removing a cake too quickly risks having parts of it stuck to the pan. While less critical, over-cooling a cake may also cause difficulty in removing it from the pan.

For foam cakes (such as sponge cakes) baked in a tube pan, invert pan to cool on a wire rack or on four glasses to retain the shape of the baked goods. Let pan stand for 1½ hours before removing the cake from it to prevent the cake from deflating. Keep rack at least two inches above the surface to avoid condensation.

3 BAKING EQUIPMENT

In addition to baking ingredients and baking process, equipment plays a very important role in baking. There is a Chinese saying, "to do a great job, you must first perfect the equipment." While experienced bakers may be able to produce quality baked goods independent of equipment, they would be able to do even better with appropriate equipment. This segment of the book discusses baking equipment according to the five steps of the baking process outlined in Chapter 2 earlier.

Please refer to photos on the opposite page

BAKING EQUIPMENT CAPTIONS

a.	Egg Separator		i.	Aluminum Foil Pan and Paper Liner
b.	Sieve		j.	Oven and Oven Rack
c.	Mixer		k.	Oven Mitts
d.	Timer		l.	Knife
e.	Freezer/Sandwich Bag		m.	Parchment Paper and Wax Paper
f.	Measuring Tools		n.	Baking Pan
g.	Spatula		o.	Wire Rack
h.	Mixing Bowl		p.	Testing Stick

a.

b.

c.

d.

e.

f.

g.

h.

i.

j.

k.

l.

m.

n.

o.

p.

INGREDIENT PREPARATION

Egg Separator

Some recipes, such as chiffon and sponge cakes, require separating the egg's white from its yolk. Some use the two cracked half eggshells to accomplish this. A more foolproof method, however, is to use an egg separator which can make the job easier and cleaner. An egg separator is a small gadget that looks like a small measuring cup. Pour an egg after it has been cracked onto the separator sitting on a container such as a wide-mouthed cup. The separator will retain the egg yolk in the middle while letting the egg white drip through it into the container. An egg separator is a handy gadget to have; it reduces the chance of breaking the egg yolk, thus contaminating the egg white with traces of fat, which is a "no no" for foam cakes.

Freezer/Sandwich Bag

A freezer or sandwich bag (plastic bag) is handy for coating fruit or nuts with a small amount of flour by shaking them in it before folding these ingredients into cake batter. The ones with a zipper help prevent spillage.

Measuring Tools

Volume measuring cups are available for measuring liquids such as juice, milk, and oil. They are commonly made of glass or plastic. For measuring dry ingredients such as flour and sugar, plastic measuring cups are commonly available in various sizes (e.g., 1 c, ½ c, ⅓ c, ¼ c, etc.). While they are the same in N. America, one dry cup is equal to 1.1636 liquid cups in Great Britain. For ingredients that cannot be compacted and have air spaces between them, e.g., dried fruit, nuts, chocolate chips, etc., weight is a more precise measurement to quantify them for reproducible results. This, however, is not always done in recipes. For measuring smaller amounts of wet (such as extracts and fruit zests) or dry ingredients (such as salt and cream of tartar), plastic or stainless steel measuring spoon sets are available for use in table (tbsp) and tea (tsp) spoon sizes (e.g., 1 tbsp/tsp, ½ tbsp/tsp, ¼ tbsp/tsp, etc.).

Sieve

For better quality baked products, sift flour to avoid lumps. Stainless steel sifters using a rotary mechanism can be purchased readily. You can also purchase a simple sieve. By tapping the side of the flour-containing sieve against the area bounded by your thumb and the index finger, you can achieve good sifting. As an alternative, you may wish to buy pre-sifted flour to save the extra work. Chemical leavener (e.g., baking powder and/or baking soda), when used, should also be sifted and mixed with flour to ensure uniform mixing to avoid possible uneven rising and large holes in baked goods.

BATTER PREPARATION

Mixer

Mixers are used for various purposes such as mixing flour and liquid(s), beating eggs, and creaming butter and sugar. Some prefer using manual hand-held mixers while most bakers use mixers operated by electricity. For small quantities of ingredients, it is best to use a hand-held mixer so that it can reach the ingredients in the bottom of the mixing bowl. For larger amounts, a mixer with a stand also works well. Mixers are available commercially in different power (wattage). The higher the power, the lesser mixing time is required. For domestic use, most hand held-mixers are not more than 300 watts. Mixers that come with a stand may have a higher wattage than those that do not, but likely not over 450 watts. Recipes that specify processing time do not typically specify mixer's power. Let us assume that the average mixer power is 200 watts. A baking recipe's specified mixing time needs to be adjusted when using a mixer with a higher or lower wattage than 200 watts to avoid over- or under-mixing or beating. When your mixer's power is lower or higher by 50 per cent, scale the recipe's specified preparation time up or down, say, by 25 per cent. For example, try increasing mixing time by 25 per cent when adjusting from 200 watts to 100 watts or decreasing mixing time by 25 per cent when adjusting from 200 watts to 300 watts. You need, however, to experiment to determine the exact required adjustment.

Mixing Bowl

Ideally, mixing bowls should be 9" to 10" in diameter and 5" to 6" deep to ensure no spillage of baking ingredients during mixing or beating. In general, a mixing bowl's material does not affect the outcome of non-foam cake baking. For foam cakes that require separation of egg whites from egg yolks; however, avoid using plastic and wooden containers as they readily absorb and retain traces of fat, thus reducing egg whites' ability to hold air bubbles. Aluminum bowls are not desirable because aluminum surfaces react with egg white, turning it slightly grey. Copper, glass, and stainless steel bowls are suitable for whipping egg whites.

Copper surface contains a natural acid that stabilizes beaten egg whites. Some copper ions from the surface migrate into the egg whites, and react with conalbumin (an egg white protein) to form a yellowish complex. This complex is more stable than conalbumin, resulting in fewer protein molecules available to denature and coagulate. Copper may also react with sulfur-containing groups from other proteins, stabilizing them. Egg whites whipped in a copper bowl are hence less likely to unfold, making the use of cream of tartar unnecessary.

Spatula

A spatula is used to scrape batter from a mixing bowl into a baking pan or to fold one batter ingredient (e.g., beaten egg yolk/flour batter) into another (e.g., beaten egg whites). It can also be used to cut through bubbles on batter's surface before baking to help reduce sudden bursts of air bubbles. Spatulas are commonly made of plastic and silicone For foam cakes, avoid using plastic and wooden spatulas because they absorb and retain fats.

Timer

In order to keep track of time throughout preparation, baking, and post-baking steps, use a timer to time recipes' processing steps. Both countdown type and sports clock type timers are commercially available.

BATTER TRANSFER

Aluminum Foil Pan and Paper Liner

Disposable aluminum foil pans and paper liners that are shaped to fit into a loaf pan are commercially available. Some find their use a convenience in baking. Without having to remove cakes from baking pans, using them helps avoid possible sticking and distortion of baked goods. Aluminum foil pans, however, absorb oven heat unevenly, thus causing possible uneven baking due to hot spots.

Baking Pan

Baking pans come in different types, shapes, dimensions, and materials. Most commonly used cake pans are loaf pans, springfoam pans, Bundt pans, and tube pans. Most common shapes are rectangular, square, and round. For some cakes, only pan type specified in recipes should be used, for instance, foam cakes such as sponge cakes need a tube pan for the batter to cling to in the centre. Pan shape, pan dimension, and pan material affect baking temperature and baking time. When substituting pan specified in baking recipes, it may be necessary to adjust baking temperature and/or baking time (see Tables 9 and 10 in Part II under "General Guidelines for Baking Pan Substitution"). You can determine the exact time required by inserting a skewer stick into the cake to test for doneness three minutes prior to the recipe's specified time.

Parchment Paper and Wax paper

Parchment paper is a non-stick paper that can be shaped as a paper lining to fit into a baking pan to prevent baked goods from sticking to the pan, thus avoiding cake breakage and deformation. Parchment paper increases the chance of success in cake removal from baking pans, preserving their perfect shapes and looks. This can be easily accomplished by cutting out and crossing two strips of parchment paper to cover the inner bottom and four sides of the pan. As an alternative, just cut out the right-sized parchment paper, fit it onto the bottom of the inside of the pan, and spray the sides of the pan with oil will do the job as well.

Like parchment paper, wax paper is another non-stick paper that can be used to form a paper lining shaped to fit into a baking pan. It reduces the risk of baked goods from sticking to the pan upon removal. Wax paper, however, does not receive oil spray as well as parchment paper and that's why some bakers prefer the use of parchment paper to wax paper.

BAKING

Oven and Oven Rack

Modern ovens are fuelled by electricity or gas. Electric ovens tend to be hotter, bake faster, and have more even heat than gas ovens. Gas ovens have moist heat and some prefer cakes baked in them to electric ovens. A convection oven differs from a conventional oven in that it has a fan inside it to distribute heat more evenly.

It is important to check out oven temperatures with an oven thermometer, especially for older ovens that might have become unreliable. Ovens may also have hot spots for which can be partially compensated by positioning a baking pan in certain parts of the oven during baking. You may need to adjust slightly the recipe-specified baking temperature and baking time to reflect oven type and age, if appropriate.

An oven rack is used to hold batter-filled pans during baking. Rack positioning is important to heat exposure of the batter and thus affect baking time. This is especially crucial for older ovens when temperature inside the oven may be uneven. Normally, place baking pan in the centre position of the middle rack.

Oven Mitts

They are a must since ovens and baking pans can get very hot. You should consider longer mitts covering your wrists and forearms to avoid getting burned due to unintentional touching of oven rack and sides.

Testing Stick

Because of temperature variability of ovens, it is desirable to check a cake's doneness three minutes prior to the recipe's specified baking time in order to avoid over-baking. Wooden sticks (used in BBQ skewing) of varying lengths are commercially available and work well for this purpose. In the absence of skewer sticks, a piece of uncooked spaghetti is a workable substitute.

POST-BAKING ACTIVITIES

Freezer Bag

In addition to being useful in coating fruit and nuts with flour, freezer bags are also good for storage of cakes for freezing purposes. Those with a zipper are especially handy.

Knife

To remove cakes from baking pans, run a blunt knife surrounding the cake inside the pan to help separate it more readily from the pan. It is important to use a knife that is long enough to reach the bottom of the pan and of such a material that it would not scratch the pan, whether the pan is made of metal or silicone.

An electric knife is recommended for slicing cakes into portions before serving. An alternative is a bread knife. Improve the quality of slicing by running hot water on the knife and drying it before use.

Wire Rack

A wire rack is used to support cakes for cooling before and after their removal from baking pans. In the case of foam cakes baked in a tube pan, as an alternative, use four glasses instead of a wire rack to support the inverted pan for cooling.

PART II:
QUICK REFERENCES

This part of the book serves as a quick reference source. It consists of a number of at-a-glance summary tables on baking ingredients, baking process, and baking equipment, as well as measurement conversions, and problems commonly encountered in baking. A one-page checklist is also included as a reminder to bakers of basic steps to avoid the risk of missing ingredients and to save time.

4 QUICK BAKING REFERENCES

INGREDIENT	WHAT	WHY
BASIC INGREDIENT		
FLOUR, e.g., all purpose, whole wheat, cake, pastry, etc.	o provides structure through protein/gluten formation and starch/gelatinization	o proteins from flour form gluten (elastic and stringy strands) upon contact with liquid(s)
	CHEMIST'S SECRETS	o gluten strands trap and hold air bubbles
	⚐ too much flour and/or mixing result in too much gluten formed, leading to dry and tough crust	o starch from flour absorbs liquid(s) to gelatinize, setting structure of baked goods
	⚐ too little flour and/or mixing do not create enough gluten for proper structure, resulting in collapsed cakes	
FLOUR SUBSTITUTE - CAKE MIXES, e.g., chocolate, yellow cake mixes, etc.	o serves as a substitute to flour	o some enjoy using cake mixes because they simplify preparation and increase the chance of success
TOUGHENERS - PRIMARY		
EGG (WHOLE)	o contributes to cakes' structure	o egg binds other ingredients together
	o also adds tender texture, colour, and flavour to cakes	
	CHEMIST'S SECRETS	
	⚐ too much egg makes baked goods soggy	
	⚐ too little egg results in baked goods being dry and lack yellowish colour	
EGG WHITE	o when whipped, forms a foamy network, giving rise to light and delicate baked products	o foam traps air
	CHEMIST'S SECRETS	o upon baking, egg protein denatures and coagulates, giving structure to the foam and air is released causing batter to rise
	⚐ too much whipped egg white traps too much air resulting in tunnels in baked products	
	⚐ too little whipped egg white does not trap enough air and cake is flat	

INGREDIENT	WHAT	WHY
TOUGHENERS — SECONDARY		
CREAM OF TARTAR	o contributes to cakes' structure	o cream of tartar stabilizes beaten egg white foam for un-shortened cakes
FAT & SUGAR	o produces good volume and fine crumb texture	o when fat is creamed with sugar, it incorporates air cells
	CHEMIST'S SECRETS	
	⚗ when over-creamed, it loses the ability to trap air	
	⚗ under-creaming results in coarse grain and lumpiness of baked goods	
LIQUID, **e.g., water**	o contributes to cakes' structure	o liquid provides moisture for gluten formation and starch gelatinization for shortened cakes
SALT	o contributes to structure of cakes	o salt strengthens glutens for shortened cakes and stabilizes beaten egg white foam for un-shortened cakes
	o provides also flavour and enhances aroma of other ingredients	
	CHEMIST'S SECRETS	
	⚗ too much salt may decrease stability of egg white foam network and result in saltiness of baked goods	
SUGAR	o contributes to cakes' texture and volume	o sugar interacts with egg proteins to stabilize beaten egg white foam structure for un-shortened cakes
	o adds sweetness and helps browning	o due to sugar's sweetness, caramelization, and Maillard reaction
	CHEMIST'S SECRETS	
	⚗ too much sugar increases process time	
	⚗ too much sugar results in sticky top of baked goods and its being stuck to pan	
	⚗ too little sugar leads to insufficient sweetness and browning	

TABLE I: *Functions of Key Ingredients At-A-Glance*

INGREDIENT	WHAT	WHY
TENDERIZERS — PRIMARY		
DAIRY PRODUCT, e.g., buttermilk, homo milk, sour cream, yogurt, etc.	o contributes to tender crust and fine texture of baked goods o adds also flavour and colour (when containing fats) to cakes CHEMIST'S SECRETS ⚠ too much dairy products makes baked goods soggy and heavy ⚠ too little dairy products may result in dry cakes	o dairy product adds moistness to cakes
EGG YOLK	o contributes to tender texture of baked goods	o egg yolk's lecithin emulsifies fats and liquids
FAT, e.g., butter, margarine, shortening, etc.	o coats flour particles o coats and shortens gluten strands o contributes to tender texture o adds also flavour to baked goods CHEMIST'S SECRETS ⚠ too much fat makes baked goods soggy and heavy ⚠ too little fat results in coarse crumb and tough crust	o reducing gluten formation and over structure o making gluten strand slippery and allowing gas bubbles to move easily o fat increases moistness o due to some fat's unique flavour, e.g., butter and margarine
LEAVENER (CHEMICAL), e.g., baking powder, baking soda	o forms and releases carbon dioxide through chemical and thermal reactions, resulting in light and porous texture, good volume, tender crumb, and uniform cells CHEMIST'S SECRETS ⚠ too much chemical leavener gives rise to large holes and dry, tough crust ⚠ too little chemical leavener forms/ releases too little carbon dioxide, resulting in poor volume and compact crumb	o carbon dioxide expands already trapped air bubbles in batter causing it to rise

TABLE I: *Functions of Key Ingredients At-A-Glance*

INGREDIENT	WHAT	WHY
LIQUID, e.g., water, juices, etc.	o evaporates at baking temperature to form steam for leavening action	o steam expands already trapped air bubbles in batter, causing it to rise and resulting in light and porous texture, good volume, tender crumb, and uniform cells
	o contributes to tender texture	o liquid increases moistness
	o adds also unique flavour	o due to juice's unique flavour
	CHEMIST'S SECRETS ⚗ too much liquid makes baked goods soggy and heavy ⚗ too little liquids may result in dry cakes	
PUDDING MIX, e.g., banana, chocolate, coconut, vanilla pudding mixes, etc.	o contributes to tender texture o introduces also flavour to cakes	o pudding mix increases moistness o due to pudding mix's unique flavour

TENDERIZERS — SECONDARY / SUPPORTING

INGREDIENT	WHAT	WHY
DAIRY PRODUCT, e.g., buttermilk, sour cream, yogurt, etc.	o contributes to rise of batter	o dairy product provides an acidic medium to react with baking soda (if used) to release carbon dioxide to leaven batter
LIQUID, e.g., fruit juice, etc.	o contributes to light and porous texture, good volume and tender crumb	o liquid, if acidic, reacts with baking soda to form and release carbon dioxide, expanding already trapped air bubbles in batter, causing it to rise
SUGAR, e.g., granulated white sugar, brown sugar, confectioner's sugar, etc.	o reduces over structure of cakes, contributing to their tenderness	o sugar competes for liquid(s) required for gluten formation and starch gelatinization

OTHER INGREDIENTS

INGREDIENT	WHAT	WHY
FLAVOURING, e.g., cocoa powder, extracts such as lemon, orange, vanilla, liqueur, etc.	o contributes to taste	o due to its unique flavour and aroma
FRUIT, e.g., fresh, frozen, dried fruit, etc.	o adds unique flavour and aroma to cakes o serves as good decoration CHEMIST'S SECRETS ⚗ too much and/or too large pieces may block expansion of air bubbles during baking, hindering baked goods to rise	o due to natural flavour and aroma of fruit

TABLE 1: *Functions of Key Ingredients At-A-Glance*

INGREDIENT	WHAT	WHY
FRUIT, **e.g., fresh, frozen, dried fruit, etc.**	▲ to cut fruit into small pieces ▲ to coat with a thin layer of flour to keep pieces separate and not sink in batter	
FRUIT ZEST, **e.g., lemon, lemon, etc.**	○ contributes to taste and mouth feel	○ due to its unique flavour and body
NUT, e.g., almond, hazelnut slices, etc.	○ adds unique flavour and mouth feel CHEMIST'S SECRETS ▲ too much and/or too large pieces may block expansion of air bubbles during baking, hindering baked goods to rise ▲ to slice nuts into small size ▲ to coat nut pieces with a thin layer of flour	○ due to nut's natural flavour

TABLE 2: *Guidance on Relative Ingredient Proportions*

INGREDIENT 1	INGREDIENT 2
FLOUR: 1 c	Baking powder: ~1 tsp (needs to scale down if baking soda is also used)
FLOUR: 1 c	Baking soda: 0 - ½ tsp
FLOUR: 1 c	Fat + Dairy Product + Liquid: ~¾ - 1½ c (depending on what other wet ingredients, e.g., mashed ripe banana, are used)
FLOUR: 1 c	Extract: ~1 tsp
FLOUR: 1 c	Sugar: ~½ to 1 c
EGG WHITES: 8	Cream of tartar: ~ ½ tsp

HIGH ALTITUDE BAKING necessitates adjustments of quantities of certain ingredients. At high altitudes, pressure is lower so is the boiling point of liquid (e.g., water). This results in faster evaporation, which causes higher sugar concentrations and accelerated expansion of air bubbles in the batter and leads to slower cake setting and popping/deflation of baked goods. You may remedy this partially by decreasing the amounts of sugar and chemical leavener (and perhaps fats) and by increasing the amounts of flour and liquid. Some baking experts have suggested guidelines but exact adjustments require empirical determination.

STAGE	OBSERVATION ON FOAM	CHEMIST'S SECRETS
FOAMY (with slight beating on low speed for about one minute*)	o is frothy, transparent, and runny o consists of large bubbles that pop readily o does not hold peaks when beaters are lifted	o add small amounts of sugar and salt to stabilize foam NOTE: o adding sugar increases beating time o too much salt breaks the foam
SOFT PEAK (with continued beating on medium low speed for about two minutes*)	o has finer division of air bubbles o is moist, shiny, and whiter o becomes stiffer, forming dull peak when beaters are lifted o flows when mixing bowl is tilted	
STIFF PEAK (with further beating on high speed for about four minutes*)	o loses moist and shiny appearance slightly o is very white, stiff, and rigid o has maximum volume and holds an upright peak when beaters are lifted o does not flow when mixing bowl is tilted	o add a small amount of cream of tartar to stabilize foam
OVER-BEATEN	o appears dry and granular o loses shiny look o films around the air bubbles are thin and non-elastic, resulting in larger air cells when ruptured o is small curd-like; white patches may appear	o add another egg white or two and continue beating in order to attempt to revert situation o discard and start over again if to no avail

* Beating time based on a 300-watt mixer.

PROCESS	PRIMARY FUNCTION	CHEMIST'S SECRETS
I INGREDIENTS PREPARATION		
I Egg preparation, beating/whipping	O to separate egg white from egg yolk O to prepare egg whites and/or egg yolks for foam cake baking	O separate when egg is cold from the fridge O allow separated egg components to return to room temperature (about 30 minutes) before beating begins
2 Fruit/nut preparation	O to prepare fruit (dry, fresh, or frozen) and nuts for folding into batter	O cut fruit into small-sized pieces ★ chill fresh fruit ★ run boiling water over dried fruit and drain ★ defrost and drain liquid from frozen fruit O slice nuts thinly or cut them into small pieces O coat fruit or nuts with a small amount of flour by shaking them in a zippered freezer/ sandwich bag to keep them separate from one another and sinking in batter
3 Pre-measuring ingredients	O to prepare ingredients (e.g., measure out and sift flour) ahead of time	O ensure all ingredients are available and ready to avoid time loss
4 Utensil cleaning	O to avoid surface contamination by cleaning utensils thoroughly	O avoid using plastic and wooden surfaces for whipping and transferring egg whites because they absorb and retain fats even after cleaning
2 BATTER PREPARATION		
I Ingredient folding	O to combine different baking ingredients together gently	O carry out step gently and steadily with a spatula to avoid over-mixing O use silicone instead of plastic or wooden spatulas for foam cakes
2 Ingredient mixing and beating	O to ensure air is trapped in batter, leading to satisfactory structure and texture of baked goods	O require longer mixing/baking time when mixing ingredients from cooler than room temperature, or vice versa O avoid over- or under-mixing/beating

TABLE 4: *Functions of Key Process Steps At–A–Glance*

PROCESS	PRIMARY FUNCTION	CHEMIST'S SECRETS
3 BATTER TRANSFER		
1 Avoiding overflow	o to avoid overflow of batter upon baking	o fill baking pan with batter to no more than 70 per cent full
2 Choosing baking pan	o to choose proper sized baking pan	o stick to recipe's recommendation if at all possible o pay attention to substituted pan's dimensions and material *(see Tables 9 and 10 in Part II)*
3 Greasing baking pan	o to apply a thin coat of butter, margarine, or oil on baking pan (or line with parchment paper as an alternative) to help easy removal of baked goods' from pan	o use oil spray to simplify process o add a thin coat of flour (or cocoa powder for chocolate cakes) on the greased surface
4 Pre-conditioning silicone baking pan	o to take action to avoid uneven heating	o preheat silicone baking pan in oven for three minutes before pouring batter into it
5 Transferring Batter	o to transfer batter from mixing bowl to baking pan	o tap batter-filled pan against counter top or use a spatula to cut through batter to release large air bubbles
4 BAKING		
1 Choosing baking temperature and baking time	o to ensure appropriate heat is available to set and bake cake	o stick to recipe's recommendations o choose baking temperature and baking time to reflect amounts of flour and other dense ingredients, and baking pan dimensions and material type, when modifying recipes
2 Pre-heating oven	o to preheat oven for 15 minutes before baking to attain even temperature	o verify oven temperature with an oven thermometer
3 Rack positioning	o to put oven rack in middle (for most recipes) or lower part (as specified in recipes) of oven	o put baking pan in the centre position of rack
4 Testing for doneness	o to test for cake's doneness about three minutes before recipe's baking time	o insert a wooden BBQ skewer stick or a piece of raw spaghetti to ensure that it comes out clean

TABLE 4: *Functions of Key Process Steps At–A–Glance*

PROCESS	PRIMARY FUNCTION	CHEMIST'S SECRETS
5 POST-BAKING		
1 Cake freezing	o to freeze cake/slices in a freezer for storage	o wrap individual cake slices with a plastic sheet before storing in a freezer bag
2 Cake slicing	o to create presentable and appealing cake slices for serving	o use an electric knife or bread knife to slice cake
		o run hot water over knife and dry before cutting
		o cut slantingly to result in larger-sized cake slices
3 Cake unmolding	o to remove baked goods from baking pan after cake has been cooled	o risk cake being stuck to baking pan when removing cake (shortened) with less than 15 minute cooling time or with much longer cooling time than 15 minutes
		o cool unshortened (foam) cakes for 1.5 hours by inverting tube pan on a wire rack or on four glasses before cake removal (keep at least 2" above surface)
		o use a blunt knife to slice around the outside of the cake inside of the pan to help release cake from pan

TABLE 5: *Functions of Key Equipment At–A–Glance*

EQUIPMENT	PRIMARY FUNCTION	CHEMIST'S SECRETS
1 INGREDIENTS PREPARATION		
1 Egg separator	o to separate egg white from egg yolk	o separate eggs when they are cold from the fridge
2 Freezer / sandwich bag	o to coat dried fruit or nuts with flour	o shake fruit/nuts and a small amount of flour in a zippered freezer /sandwich bag
3 Measuring tools, e.g., dry and liquid measuring cups, spoon sets, weighing scale	o to measure ingredients, by volume or by weight	o do not compact dry ingredients while measuring
		o measure loose, dry ingredients (e.g., dried fruit) by weight (e.g., grams), instead of by volume (e.g., cup)

TABLE 5: *Functions of key Process Steps At–A–Glance*

	EQUIPMENT	PRIMARY FUNCTION	CHEMIST'S SECRETS
4	Sieve	o to sift flour and powdered leavener to avoid lumps and to ensure uniformity	o make sure sieve is dry o use sieve of appropriate pore size, i.e., not too small
2 BATTER PREPARATION			
1	Mixer and mixer accessories, e.g., beater, whisk, etc.	o to mix or beat ingredients using different accessories	o adjust mixing time to reflect mixer's power (i.e., wattage; see *Chapter 3 "mixer" under "Batter Preparation"*)
2	Mixing bowl	o to contain baking ingredients for mixing and beating	o use mixing bowls that are wide and deep enough to avoid spillage o avoid using plastic and wooden surfaced mixing bowls for foam cakes as they absorb and retain fats
3	Spatula	o to fold one ingredient into another, to transfer batter into baking pan, and to cut across batter to release large air bubbles	o do not use plastic or wooden spatulas for foam cakes because they absorb and retain fats
4	Timer	o to time preparation, baking, and post-baking steps	o use a countdown timer with a buzzer to simplify timing procedure
3 BATTER TRANSFER			
1	Aluminum foil pan and paper liner	o to contain batter for baking to reduce finished baked goods from sticking to baking pan	o note that aluminum pans may have hotbspots causing possible uneven heating
2	Baking pan, e.g., Bundt pan, loaf pan, springfoam pan, tube pan, etc.	o to house batter to shape baked goods o to provide support for some cakes to rise (e.g., tube pan for sponge cake)	o need to adjust baking temperature and/or baking time to reflect colour, shape, size, and material of baking pans
3	Parchment paper and wax paper	o to use as non-stick paper to avoid baked goods from sticking to baking pans	o spray paper with oil to doubly ensure that baked goods are not stuck to pan
4 BAKING			
1	Oven, e.g., electric oven, gas oven and oven rack	o to provide heat to cause batter to rise and bake into final products o to support baking pan	o pre-heat oven for 15 minutes o check oven temperature accuracy and evenness with an oven thermometer o put cake in the centre position of middle rack, unless specified by recipe

TABLE 5: *Functions of key Process Steps At–A–Glance*

	EQUIPMENT	PRIMARY FUNCTION	CHEMIST'S SECRETS
2	Oven mitts	○ To remove baking pans from oven	○ use long mitts to avoid getting burned by oven's sides and racks unintentionally
3	Testing stick	○ to test for doneness of baked goods	○ use wooden BBQ skewer sticks commercially available in different lengths
			○ use an uncooked spaghetti as an alternative substitute
5 POST-BAKING			
1	Freezer bag	○ to store cake slices in a freezer for future consumption	○ wrap individual slices with a plastic wrap before storing them in a zippered freezer bag
2	Knife, e.g., bread knife, electric knife	○ to loosen cake from baking pan	○ use a blunt knife to help remove cake from pan
		○ to slice cake for serving	○ run hot water over an electric or a bread knife and dry before slicing
3	Wire rack	○ to support baked goods to cool	○ invert and keep foam cakes at least 2" from surface to avoid condensation

TABLE 6: *Common Liquid (or Volume) Measurements*

ONE	APPROXIMATE EQUIVALENT
TEASPOON (tsp)	60 drops
	⅓ tbsp
	⅙ fl oz.
	5 ml

ONE	APPROXIMATE EQUIVALENT
TABLESPOON (tbsp)	3 tsp
	½ fl oz.
	¹⁄₁₆ c
	15 ml
FLUID OUNCE (fl oz.)	6 tsp
	2 tbsp
	⅛ c
	30 ml
CUP (c)	48 tsp
	16 tbsp
	8 fl oz.
	250 ml
PINT (pt)	32 tbsp
	2 c
	16 fl oz.
	500 ml
QUART (qt)	2 pt
	4 c
	32 fl oz.
	1000 ml = 1 L
GALLON (gal)	4 qt
	8 pt
	4 L

TABLE 6: *Common Liquid (or Volume) Measurements*

ONE	APPROXIMATE EQUIVALENT
⅛ tsp	0.5 ml
¼ tsp	1 ml
½ tsp	2.5 ml
¾ tsp	4 ml
1 tsp	5 ml
1 tbsp	15 ml

ONE	APPROXIMATE EQUIVALENT
1/16 c	1 tbsp
⅛ c	2 tbsp
⅙ c	2 tbsp + 2 tsp
¼ c	4 tbsp
⅓ c	5 tbsp + 1 tsp
⅜ c	6 tbsp
½ c	8 tbsp
⅔ c	10 tbsp + 2 tsp
¾ c	12 tbsp
1 c	16 tbsp (=48 tsp)

There is no simple conversion for dry ingredients from volume (e.g., cup) to weight (e.g., grams). It depends on the density of the ingredient according to: mass (m) = volume (v) × density (d). The denser the material, the heavier the weight is. The following table lists information on a number of commonly used baking ingredients.

ONE	EQUAL TO
1 cup	16 tbsp
½ cup	8 tbsp
⅛ cup	2 tbsp
1 tbsp	3 tsp

ONE CUP	APPROXIMATE EQUIVALENT (1 oz. = 28.38g)
All Purpose Flour	4.5 oz. = 130 g
Cake Flour	3.9 oz. = 110 g
Pastry Flour	4.25 oz. = 120 g
Whole Wheat Flour	4.25 oz. = 120 g
Cornmeal, Fine	6.3 oz. = 180 g
Cornmeal, Coarse	4.85 oz. = 140 g
Chocolate Chips	5.35 oz. = 150 g
Walnut, Chopped	4.25 oz. = 120 g
Walnut / Pecan Halves	3.5 oz. = 100 g
Apricot, Dried	4.25 oz. = 120 g
Cranberry, Dried	4.25 oz. = 120 g
Coconut, Dried, Shredded	2.5 oz. = 70 g

TABLE 8: *Common Imperial and Metric Conversions*

TEMPERATURE (C = Centigrade; F = Fahrenheit) $°F = (9/5 \times °C) + 32; °C = (°F - 32) \times 5/9$	
DEGREE F	**DEGREE C** (ROUNDED OFF)
300	150
325	160
350	180
375	190
400	205
425	220

DRY WEIGHT	
ONE	**EQUAL TO** (ROUNDED OFF)
Kilogram (kg)	2.2 lb
Ounce (oz.)	30 g
8 oz.	225 g
16 oz. = 1 pound (lb)	450 g

LIQUID VOLUME	
ONE	**EQUAL TO** (ROUNDED OFF)
Fluid Ounce (fl oz.)	30 ml
4 fl oz.	125 ml = ½ cup (c)
8 fl oz.	250 ml = 1 c
16 fl oz.	500 = 2 c

LENGTH	
ONE	**EQUAL TO** (ROUNDED OFF)
Inch (in)	2.5 cm
Centimetre (cm)	10 mm

GENERAL GUIDELINES FOR BAKING PAN SUBSTITUTION; TABLES 9 AND 10

Ideally, a baker should use the type of baking pan called for in a cake recipe. Substitution by the wrong sized baking pan risks batter overflow, burned edges and bottom, and a collapsed centre. The following two tables, however, are included to provide some guidance when you must make a baking pan substitution.

1 Substitute a baking pan by one that is of similar volume or by a number of baking pans of which the sum of their volumes is similar.

2 Adjust baking time by pan dimensions:
   i   use lesser baking time when substituting with a shallower but wider pan
   ii use longer baking time when substituting with a deeper but narrower pan

3 Use slightly shorter baking time (say, five to 10 minutes) when using a silicone baking pan.

4 Adjust baking temperature by pan types:
   i   use a lower baking temperature (say, by 25°F) when substituting a shiny metal pan with a glass or a dark, non-stick pan.
   ii use a higher baking temperature (say, by 25°F) when substituting a glass or a dark, non-stick pan with a shiny metal pan.

TABLE 9: *Baking Pan Volumes*

	ONE		EQUAL TO (ROUNDED OFF)
LOAF PAN	5¾" x 3⅛" x 2"	14.38 cm x 7.81 cm x 5 cm	0.56 litre (0.5 L)
	8" x 4" x 3"	20 cm x 10 cm x 7.5 cm	1.5 litres (1.5 L)
	8½" x 4½" x 2½"	21.25 cm x 11.25 cm x 6.25 cm	1.49 litres (1.5 L)
	9" x 5" x 3"	22.5 cm x 12.5 cm x 7.5 cm	2.11 litres (2.0 L)
RECTANGULAR PAN	10¾" x 7" x 1½"	26.88 cm x 17.5 cm x 3.75 cm	1.76 litres (2.0 L)
	12" x 8" x 2"	30 cm x 20 cm x 5 cm	3.00 litres (3.0 L)
	13" x 9" x 2"	32.5 cm x 22.5 cm x 5 cm	3.66 litres (3.5 L)
ROUND PAN	8" x 1½"	20 cm x 3.75 cm	1.18 litres (1.0 L)
	8" x 2"	20 cm x 5 cm	1.57 litres (1.5 L)
	9" x 1½"	22.5 cm x 3.75 cm	1.49 litres (1.5 L)
	9" x 2"	22.5 cm x 5 cm	1.99 litres (2.0 L)

TABLE 9: *Baking Pan Volumes*

	ONE		EQUAL TO (ROUNDED OFF)
SQUARE PAN	8" × 8" × 1½"	20 cm × 20 cm × 3.75 cm	1.50 litres (1.5 L)
	8" × 8" × 2"	20 cm × 20 cm × 5 cm	2.00 litres (2.0 L)
	9" × 9" × 1½"	22.5 cm × 22.5 cm × 3.75 cm	1.90 litres (2.0 L)
	9" × 9" × 2"	22.5 cm × 22.5 cm × 5cm	2.53 litres (2.5 L)
SPRINGFOAM PAN	7¼" × 3½"	18.125 cm × 8.75 cm	2.26 litres (2.5 L)
	8" × 3"	20 cm × 7.5 cm	2.36 litres (2.5 L)
	8¼" × 2½"	20.625 cm × 6.25 cm	2.09 litres (2.0 L)
	9½" × 2½"	23.75 cm × 6.25 cm	2.77 litres (3.0 L)
BUNDT PAN	8½" × 4¼"	21.25 cm × 10.625 cm	3.71 litres (3.5 L)
	9½" × 3¼"	23.75 cm × 8.125 cm	3.36 litres (3.5 L)
	10" × 3¾"	25 cm × 9.4 cm	4.61 litres (4.5 L)
TUBE PAN	9½" × 4"	23.75 cm × 10 cm	4.24 litres (4.0 L)
	9" × 4½"	22.5 cm × 11.25 cm	4.31 litres (4.5 L)
	10" × 4"	25 cm × 10 cm	4.91 litres (5.0 L)

PAN VOLUME	DIMENSIONS		PAN TYPE
0.5L	5¾" × 3⅛" × 2"	14.38 cm × 7.81 cm × 5 cm	loaf pan
1L	8" × 1½"	20 cm × 3.75 cm	round pan
1.5 L	8" × 4" × 3"	20 cm × 10 cm × 7.5 cm	loaf pan
	8½" × 4½" × 2½"	21.25 cm × 11.25 cm × 6.25 cm	loaf pan
	8" × 2"	20 cm × 5 cm	round pan
	9" × 1½"	22.5 cm × 3.75 cm	round pan
	8" × 8" × 1½"	20 cm × 20 cm × 3.75 cm	square pan
2 L	9" × 5" × 3"	22.5 cm × 12.5 cm × 7.5 cm	loaf pan
	9" × 2"	22.5 cm × 5 cm	round pan
	8" × 8" × 2"	20 cm × 20 cm × 5 cm	square pan
	10¾" × 7" × 1½"	26.88 cm × 17.5 cm × 3.75 cm	rectangular pan
	9" × 9" × 1½"	22.5 cm × 22.5 cm × 3.75 cm	square pan
	8¼" × 2½"	20.625 cm × 6.25 cm	springfoam pan
2.5 L	9" × 9" × 2"	22.5 cm × 22.5 cm × 5 cm	square pan
	7¼" × 3½"	18.125 cm × 8.75 cm	springfoam pan
	8" × 3"	20 cm × 7.5 cm	springfoam pan
3 L	12" × 8" × 2"	30 cm × 20 cm × 5 cm	rectangular pan
	9½" × 2½"	23.75 cm × 6.25 cm	springfoam pan
3.5 L	13" × 9" × 2"	32.5 cm × 22.5 cm × 5 cm	rectangular pan
	9½" × 3¼"	23.75 cm × 8.125 cm	Bundt pan
4 L	8" × 5"	20 cm × 12.5 cm	springfoam pan
	9½" × 4"	23.75 cm × 10 cm	tube pan
4.5L	10" × 3¾"	25 cm × 9.4 cm	Bundt pan
	9" × 4½"	22.5 cm × 11.25 cm	tube pan
5L	10" × 4"	25 cm × 10 cm	tube pan

5 FREQUENTLY ENCOUNTERED PROBLEMS

The following table lists many common baking problems. Typically, the reason for a baking problem occurrence is due not to a single but multiple causes. The likely causes are presented together with suggested adjustments.

TABLE 11: *Common Baking Problems*

OBSERVED PROBLEMS	POSSIBLE CAUSE	CHEMIST'S SECRETS
I STRUCTURE		
I Cake batter overflows	o too much baking powder/soda	o use less baking powder/soda
	o too much batter in baking pan or pan size too small	o fill with batter to no more than 70 per cent of baking pan or use larger-sized pan
	o heat too high	o decrease baking temperature
2 Cake burned on outside but inside is undone	o oven heat too high and not enough baking time	o lower baking temperature and increase baking time
3 Cake lacks body	o not enough flour and/or under-mixed	o use more flour and/or increase mixing time
	o too much fat/liquid	o decrease amounts of fat/liquid
	o too much sugar	o use less sugar
4 Cake peaks and cracks on top	o batter over-mixed	o reduce mixer speed and/or mixing time
	o oven too hot and/or over baked	o lower baking temperature and/or reduce baking time
	o too much baking powder/soda	o use less baking powder/soda
5 Cake rises poorly	o not enough baking powder/soda	o use more baking powder/soda
	o oven heat too low	o increase baking temperature
	o too much fat or liquid	o use less fat or liquid

OBSERVED PROBLEMS	POSSIBLE CAUSE	CHEMIST'S SECRETS
6 Cake rises and then caves in	o too much baking powder/soda	o reduce amount of baking powder/soda
	o batter over-mixed	o decrease mixing time and/or mixer speed
	o oven heat too high	o use lower baking temperature
	o too much liquid	o reduce amount of liquid
7 Cake rises unevenly	o uneven mixing of baking powder/soda	o sift and mix baking powder/soda with flour
	o uneven heat inside oven	o pre-heat oven for 15 minutes and avoid opening oven door while baking
8 Cake sticks to pan	o inadequate greasing of baking pan	o properly grease pan and coat with a small amount of flour (or line with parchment paper)
	o removing cake from pan too soon or too late after cooling	o let cake cool in pan on a wire rack for 15 minutes prior to removal
	o too much sugar	o reduce amount of sugar
9 Fruit and nuts sink to bottom of cake batter	o fruit/nut pieces too heavy	o cut fruit/nuts to smaller size
		o mix and coat fruit/nuts with a small amount of flour

2 TEXTURE		
1 Cake has bitter taste	o too much baking soda/powder	o decrease amount of baking powder/soda
2 Cake has coarse grain	o under-creaming of fat/sugar	o cream fat/sugar more
	o batter under-mixed	o increase batter mixing time
	o too much oil	o use less oil
3 Cake is dense/heavy	o too much flour or over-mixing	o use less flour and/or mixing time
	o too much fat/liquid	o use less fat/liquid
	o oven heat too low	o increase baking temperature
4 Cake is dry with tough crust	o oven too hot and/or baked too long	o decrease baking temperature and/or baking time
	o too much flour	o reduce amount of flour
	o not enough fat/liquid	o increase amount of fat/liquid
	o too much baking powder/soda	o use less baking powder/soda

TABLE II: *Common Baking Problems*

OBSERVED PROBLEMS	POSSIBLE CAUSE	CHEMIST'S SECRETS
5 Cake has large holes (tunnelling)	o too much and/or uneven mixing of baking powder/soda	o decrease amount and/or ensure uniform mixing of baking powder/soda
	o batter over-mixed	o reduce mixing time
	o oven too hot	o decrease baking temperature
6 Cake is lumpy	o under-creaming of fat/sugar	o proper creaming of fat and sugar
	o uneven mixing of ingredients	o mix batter longer
7 Cake does not have enough browning	o baking time too short and/or oven heat too low	o increase baking time and/or baking temperature
	o not enough sugar	o add more sugar
8 Cake has sticky top	o cake covered when still warm, causing condensation	o do not cover cake until cake is cooled completely
	o too much sugar	o decrease amount of sugar
9 Cake is soggy	o too much fat and/or liquid	o use less fat/liquid
	o too many eggs and/or eggs under-beaten	o decrease number of eggs and/or beat eggs more

6 A BAKER'S CHECKLIST

INGREDIENT PREPARATION

- [ ] Read and familiarize with recipe's ingredients and steps
- [ ] Make sure all required ingredients and equipment are available
- [ ] Separate eggs (when required by recipe) when cold from the fridge
- [ ] Leave eggs for at least 30 minutes in room temperature before beating eggs
- [ ] Test baking powder and baking soda for efficacy
- [ ] Sift flour, baking powder, and baking soda (if used)
- [ ] Pre-measure all ingredients

BATTER PREPARATION

- [ ] Do not over- or under- mix batter (i.e., follow recipe instructions)
- [ ] Do not use plastic or wooden spatulas and mixing bowls for foam cakes

BATTER TRANSFER

- [ ] Use recipe-specified baking pan (see Tables 9 and 10 if substitution is necessary)
- [ ] Grease/spray with oil and flour baking pan (or line pan with parchment paper)
- [ ] Pre-heat silicone pan for 3 minutes (if used in baking)
- [ ] Fill baking pan with batter to no more than 70 per cent full
- [ ] Tap filled pan against counter or cut batter with a spatula to release large air bubbles

BAKING

- [ ] Place oven rack in middle level unless specified in recipe
- [ ] Pre-heat oven to recipe-specified temperature for 15 minutes
- [ ] Check baked goods' doneness three (3) minutes prior to the recipe's specified baking time with a wooden skewer stick

POST–BAKING

- [ ] Set cake pan (for shortened cakes) on a wire rack for 15 minutes to cool
- [ ] Invert tube cake pan (for foam cakes) on a wire rack or on four glasses (at least 2" from the surface) for 1½ hours to cool
- [ ] Run a blunt knife around the cake inside of pan to loosen cake
- [ ] Slice cake with an electric or a bread knife
- [ ] Enjoy the fruit of your labour!!!

PART III:
BAKING YOUR CAKE LIKE A CHEMIST

This part of the book includes 12 selected recipes that I have personally developed and tested.

Why only 12 recipes? For most people who entertain friends with desserts, they do it mostly over the weekends. Most of the times, these parties involve different people. Even if the same guests are invited once per month, you will have a different quality cake for them every month for a whole year. Therefore, having 12 recipes that you can have confidence in is more than adequate.

When you target to try out one recipe every month to develop your expertise, you will master 12 good recipes in a year. For those who are more ambitious, make it a 12-week plan. Furthermore, with a good understanding of the baking framework described in Part I, you will be able to modify the basic recipes to create your own new recipes, thus turning the 12 recipes to become many more recipes.

Each recipe is accompanied by a photograph and a brief description of the cake. In each recipe, baking ingredients have been standardized and steps quantified to remove uncertainties. Time determinations in these recipes are based on the use of a 300 watt mixer. For mixers with a scale of 0 to 6: 0=off, 1=low, 2=medium low, 3=medium, 4=medium high, 5=high, and 6=very high. The preparation times specified in the recipes would require scaling up or down if your mixer has a lower or higher wattage (see guidance in Chapter 3 on "Mixer" under "Batter Preparation").

Unique to this collection, pre-preparation steps are separated from actual preparation steps. For each recipe, baking ingredients are listed according to the sequence when they are used. They are shown in a simplified baking framework diagram as to how they relate to the "basic ingredient", "tougheners", and "tenderizers". Those ingredients that are not shown explicitly in the individual diagrams are "other ingredients". In addition, nutritional values and chemist's secrets are included for each cake to make the recipes more informative and useful.

Use all ingredients at room temperature. Increase the preparation times specified in the recipes if ingredients you use are at a colder temperature and vice versa.

the ABC CAKE

A = Apricot; **B** = Banana; & **C** = Cranberry

This cake is simple to make but the return to your investment is huge. Everyone who has tasted this cake has nothing but amazing praise to say about this cake. Because of the beautiful, appealing colours of apricot and cranberry contrasting mashed banana's lighter background colour, you can also serve this as a welcome, light fruit cake in Christmas.

■ Flour

□ Tougheners (butter & sugar, egg, salt, yogurt)

■ Tenderizers (baking powder/soda, banana, butter, sugar, yogurt)

SERVINGS PER RECIPE: 50
NUTRITIONAL INFORMATION PER SERVING

		% D.V.*
Calories	78.3	3.7
Total Fat	2.7 g	4.1
Sat. Fat	1.6 g	7.8
Cholesterol	19.0 mg	6.3
Total Carbs	11.7 g	3.9
Sugars	6.8 g	
Diet. Fibres	0.5 g	2.2
Protein	1.2 g	2.4
Sodium	31.9 mg	1.3
Calcium	18.6 mg	1.9
Potassium	62.5 mg	1.8

*%D.V. = per cent daily values are based on a 2,000 calorie diet. Your D.V. may be higher or lower depending on your calorie needs.

CHEMIST'S SECRETS

▲ Run 1½ c boiling water over dried (cut-up) apricot and cranberry and drain for 10 minutes. Coat drained fruit with a thin layer of flour by shaking them in a zippered freezer/sandwich bag.

▲ Apricot and cranberry can be substituted with other fruit (e.g., dried apple) and/or nuts (e.g., walnut). It is only limited by your imagination.

▲ If fresh bananas are not available, frozen bananas can be used. Defrost and wipe off condensation before unpeeling and mashing them.

▲ Sour cream can be used to replace yogurt if yogurt is unavailable.

PRE-PREPARATION

1 Sift and mix all purpose flour, cake flour, baking powder, and baking soda in a medium-sized bowl and put aside.

2 Cut dried apricot and dried cranberry into small pieces. Coat with flour and put aside.

3 Mash ripe banana in a medium-sized bowl and put aside.

4 Grease and flour (or line with parchment paper) five (5) 3"x6"x2" loaf pans.

5 Set oven to 350 °F.

INGREDIENTS

1¼ c	all purpose flour
¾ c	cake flour
1¾ tsp	baking powder
¼ tsp	baking soda
90 g	dried apricot
90 g	dried cranberry
1½ c	mashed ripe banana
⅗ c	butter (unsalted)
¾ c + 1 tbsp	sugar
⅛ tsp	salt
3	large eggs
1½ tsp	vanilla extract
1 c	vanilla yogurt (1.5%)

PREPARATION STEPS

1 In a large-sized bowl, beat shaved butter for 30 seconds on medium speed.

2 Add sugar and salt and cream on medium speed for 1½ minutes.

3 Add eggs and vanilla extract to butter mixture and beat for 1 minute on medium speed.

4 Add yogurt, mashed banana, and flour mixture.

5 Blend with a spatula and then mix well for 2 minutes on medium low speed.

6 Mix in cut-up dried apricot and cranberry for 1 minute on low speed.

7 Pour batter into prepared pans and bake for 45 minutes or until a skewer stick inserted into the centre of the cake comes out clean.

8 Cool cakes on a wire rack for 15 minutes before removing cakes from pans.

9 Enjoy!

almond
POUND CAKE

This cake has a strong buttery taste. The cake's texture is firm but moist and is quite filling. The cake has a creamy yellow colour inside and brown colour on the outside. This, together with sliced almonds' brown colour edges, creates a contrast that is pleasing to the eyes.

CHEMIST'S SECRETS

🧪 Use yogurt if sour cream is not available.

🧪 For decorative purposes, spread some almond slices in the bottom of the Bundt cake pan before pouring batter into it.

🧪 As an alternative, use blueberries instead of sliced almond in recipe.

PRE-PREPARATION

1 Sift all purpose flour, baking powder, and baking soda into a medium-sized bowl and set aside.

2 Weigh out 40 g of sliced almond and put aside.

3 Grease and flour a 10" Bundt cake pan

4 Set oven to 350 °F.

■ Flour
☐ Tougheners (butter/shortening & sugar, egg, milk, salt, sour cream)
■ Tenderizers (baking powder/soda, butter, milk, shortening, sour cream, sugar)

SERVINGS PER RECIPE: 24
NUTRITIONAL INFORMATION PER SERVING

		% D.V.*
Calories	240.0	12.0
Total Fat	12.2 g	18.7
Sat. Fat	6.7 g	33.5
Cholesterol	68.9 mg	23.0
Total Carbs	29.7 g	9.9
Sugars	17.2 g	
Diet. Fibres	0.5 g	1.8
Protein	3.6 g	7.1
Sodium	112.3 mg	4.7
Calcium	60.6 mg	6.1
Potassium	51.3 mg	1.5

%D.V. = per cent daily values are based on a 2,000 calorie diet. Your D.V. may be higher or lower depending on your calorie needs.

INGREDIENTS

3 c	all purpose flour
2½ tsp	baking powder
½ tsp	baking soda
40 g	sliced un-blanched almond
1 c	butter (unsalted)
¼ c	shortening
2 c	sugar
¼ tsp	salt
5	large eggs
1 tsp	almond extract
1 tsp	vanilla extract
¾ c	milk
¾ c	sour cream (1.5%)

PREPARATION STEPS

1 In a large-sized bowl, beat shaved butter and shortening for 1 minute on medium speed.

2 Add sugar and salt and cream on medium speed for 2 minutes.

3 Add eggs, almond extract, and vanilla extract. Beat on medium speed for 1½ minutes.

4 Add milk, sour cream, and flour mixture to above mixture.

5 Blend with a spatula. Then mix well for 2½ minutes on medium low speed.

6 Fold sliced almond into batter gently with a spatula.

7 Pour batter into prepared pan.

8 Bake for 60 minutes or until a skewer stick inserted into the centre of the cake comes out clean.

9 Cool on a wire rack for 15 minutes before removing cake from pan.

10 Enjoy!

angel FOOD CAKE

■ Flour

□ Tougheners (cream of tartar, egg white, salt, sugar)

■ Tenderizers (sugar)

This cake uses egg whites only (i.e., no egg yolks) and is especially welcome by those who appreciate a cake with a sponge-like texture but not cholesterol. This cake does not rise as much as a chiffon cake or a sponge cake because baking powder is not used. Lightly toast cake slices and spread butter, margarine, or jelly on it before eating.

CHEMIST'S SECRETS

Separate egg white from egg yolk while egg is cold from the fridge.

Start preparation steps only when egg whites are at room temperature (about 30 minutes after separation).

Superfine sugar can be made from granulated sugar using a coffee bean grinder or a blender.

Avoid using plastic or wooden mixing bowls and spatulas as they absorb and retain fat, which is a "no no" for foam cakes.

PRE-PREPARATION

1   Separate egg whites from egg yolks. Retain egg whites in a large-sized bowl for later use.

2   Sift flour into a medium-sized bowl and set aside.

3   Set oven to 325 °F.

SERVINGS PER RECIPE: **20**
NUTRITIONAL INFORMATION PER SERVING

		% D.V.*
Calories	64.8	3.2
Total Fat	0.1 g	0.2
Sat. Fat	0 g	0
Cholesterol	0 mg	0
Total Carbs	13.5 g	4.5
Sugars	7.7 g	
Diet. Fibres	0.3 g	1.0
Protein	2.9 g	5.8
Sodium	47.6 mg	2.0
Calcium	13.9 mg	1.4
Potassium	57.2 mg	1.6

*%D.V. = per cent daily values are based on a 2,000 calorie diet. Your D.V. may be higher or lower depending on your calorie needs.

INGREDIENTS

12	egg whites **(from large eggs)**
1¼ c	cake flour
¾ c	superfine sugar
⅛ tsp	salt
1 tsp	cream of tartar
½ tsp	almond extract
1½ tsp	vanilla extract

PREPARATION STEPS

1 Add superfine sugar (1 tbsp) and salt to egg whites. Beat for 1 minute on low speed and for another 2 minutes on medium low speed.

2 Add cream of tartar, almond extract, and vanilla extract to egg white mixture and beat for 4 minutes on high speed.

3 Add superfine sugar (¾ c minus 1 tbsp) to above mixture and beat on medium high speed for 1 more minute.

4 Sift and fold flour into above mixture with a silicone spatula over 1 minute.

5 Pour final mixture into an un-greased 10" tube pan.

6 Bake for 35 minutes or until a skewer stick inserted into the centre of the cake comes out clean.

7 Invert and place cake pan on a wire rack (or rest on 4 glasses) to cool. Remove cake from pan after 1½ hours.

8 Enjoy!

banana CARROT CAKE

Although this cake is simple to make, it is a welcome cake because of its unique texture and aroma. It has the moist texture of a banana cake as well as the flavour of a carrot cake.

■ Flour
□ Tougheners (butter & sugar, egg, salt)
■ Tenderizers (baking powder/soda, banana, butter, carrot, sugar)

CHEMIST'S SECRETS

🧪 Add 90 g of chopped walnut or pecan as preparation step #6 (and as a new #4 in pre-preparation step), if desired.

🧪 Coat walnut or pecan (if used) with a thin layer of flour by shaking them in a zippered freezer/sandwich bag.

🧪 If fresh bananas are not available, frozen bananas can be used. Defrost and wipe off condensation before unpeeling and mashing them.

PRE-PREPARATION

1 Sift all purpose flour, baking powder, and baking soda into a medium-sized bowl and set aside.

2 Mash ripe banana in a medium-sized bowl and put aside.

3 Grate carrot and put aside.

4 Grease and flour (or line with parchment paper) four (4) 3"x6"x2" loaf pans.

5 Set oven to 350 °F.

SERVINGS PER RECIPE: 40
NUTRITIONAL INFORMATION PER SERVING

		% D.V.*
Calories	72.0	3.6
Total Fat	3.2 g	4.9
Sat. Fat	1.9 g	9.4
Cholesterol	23.2 mg	7.7
Total Carbs	9.8 g	3.3
Sugars	4.9 g	
Diet. Fibres	0.4 g	1.4
Protein	1.2 g	2.3
Sodium	43.3 mg	1.8
Calcium	15.8 mg	1.6
Potassium	35.5 mg	1.0

%D.V. = per cent daily values are based on a 2,000 calorie diet. Your D.V. may be higher or lower depending on your calorie needs.

INGREDIENTS

1 ¾ c	all purpose flour
1 ½ tsp	baking powder
¼ tsp	baking soda
1 ¼ c	mashed ripe banana
¾ c	grated carrot
⅗ c	butter (unsalted)
¾ c + 1 tbsp	sugar
¼ tsp	salt
3	large eggs
1 ½ tsp	vanilla extract

PREPARATION STEPS

1 In a large-sized bowl, beat shaved butter for 30 seconds on medium speed.

2 Add sugar and salt and cream on medium speed for 1 ½ minutes.

3 Add eggs and vanilla extract. Beat on medium speed for 1 minute.

4 Add mashed banana and flour mixture to mixture.

5 Blend with a spatula and then mix well for 2 minutes on medium low speed.

6 Mix in grated carrot for 1 minute on low speed.

7 Pour batter into prepared pans and bake for 45 minutes or until a skewer stick inserted into the centre of the cake comes out clean. (Alternatively: use one 5"x 9"x 3" and one 3"x 6"x 2" pans; add 10 minutes baking time for the former.)

8 Cool cakes on a wire rack for 15 minutes before removing cakes from pans.

9 Enjoy!

butter yogurt
WALNUTCAKE

This cake is simple to make and has a unique taste. Lightly toasting cake slice(s) brings out a unique flavour of the cake. For those with a sweet tooth, spread a thin layer of jelly on the toasted slice(s) before eating.

■ Flour
□ Tougheners (butter & sugar, egg, salt, yogurt)
■ Tenderizers (baking powder/soda, butter, sugar, yogurt)

CHEMIST'S SECRETS

🧪 Coat walnut with a thin layer of flour by shaking them in a zippered freezer/sandwich bag.

🧪 If walnut is unavailable, substitute it with pecan.

🧪 Use sour cream to replace yogurt if it is unavailable.

PRE-PREPARATION

1 Sift and mix cake flour, baking powder, and baking soda and set aside in a medium-sized bowl.

2 Chop walnut, coat with flour, and set aside.

3 Grease and flour (or line with parchment paper) three (3) 3"x6"x2" loaf pans.

4 Set oven to 325 °F.

SERVINGS PER RECIPE: 30
NUTRITIONAL INFORMATION PER SERVING

		% D.V.*
Calories	111.2	5.6
Total Fat	7.1 g	10.9
Sat. Fat	0.8 g	3.8
Cholesterol	33.9 mg	11.3
Total Carbs	11.0 g	3.7
Sugars	6.4 g	
Diet. Fibres	0.5 g	2.0
Protein	7.8 g	15.6
Sodium	52.0 mg	2.2
Calcium	35.7 mg	3.6
Potassium	38.8 mg	1.1

*%D.V. = per cent daily values are based on a 2,000 calorie diet. Your D.V. may be higher or lower depending on your calorie needs.

INGREDIENTS

1½ c	cake flour
1½ tsp	baking powder
¼ tsp	baking soda
90 g	chopped walnut
¾ c	butter (unsalted)
¾ c + 1 tbsp	sugar
⅛ tsp	salt
3	large eggs
1½ tsp	vanilla extract
1 c	vanilla yogurt (1.5%)

PREPARATION STEPS

1 In a large bowl, beat shaved butter for 45 seconds on medium speed.

2 Add sugar and salt and cream on medium speed for 1½ minutes.

3 Add eggs and vanilla extract. Beat on medium speed for 1 minute.

4 Add vanilla yogurt and flour mixture to butter mixture.

5 Blend with a spatula and then mix well for 1½ minutes on medium low speed.

6 Add chopped walnut (80 g) to mixture and mix on low speed for 1 minute.

7 Pour batter into prepared pans and spread remaining chopped walnut on top of batter.

8 Bake for 45 minutes or until a skewer stick inserted into the centre of the cake comes out clean. (Alternatively: use one 5"x 9"x 3" loaf pan and bake for 55 minutes.)

9 Cool cakes on a wire rack for 15 minutes before removing them from pans.

10 Enjoy!

carrot pineapple
COCONUT CAKE

This cake is easy to make and it is very moist and very tasty. The cake's taste is unique compared to the typical carrot cake. The additions of crushed pineapple, shredded coconut, and some cinnamon enhance the cake's flavour but not overpowering.

■ Flour

□ Tougheners (butter & sugar, egg, milk, salt)

■ Tenderizers (baking powder/soda, butter, carrot, milk, pineapple, sugar)

CHEMIST'S SECRETS

If unavailable, use pecan in place of walnut.

For those who wish a stronger spicy aroma, use 1½ to 2 tsp (instead of 1 tsp) of cinnamon.

PRE-PREPARATION

1   Sift and mix all purpose flour, baking powder, and baking soda in a medium-sized bowl.

2   Drain crushed pineapple from can for 10 minutes and keep in a small-sized bowl.

3   Grate carrot into a medium-sized bowl.

4   Weigh and chop walnut and put aside.

5   Weigh shredded coconut and put aside.

6   Grease and flour (or line with parchment paper) one 9"x13"x2" rectangular pan.

7   Set oven to 350 °F.

SERVINGS PER RECIPE: 36
NUTRITIONAL INFORMATION PER SERVING

		% D.V.*
Calories	122.0	6.1
Total Fat	8.8 g	13.7
Sat. Fat	4.5 g	22.2
Cholesterol	37.3 mg	12.2
Total Carbs	8.9 g	3.0
Sugars	2.3 g	
Diet. Fibres	1.0 g	4.0
Protein	2.6 g	5.1
Sodium	74.7 mg	3.1
Calcium	28.4 mg	2.8
Potassium	61.1 mg	1.8

%D.V. = per cent daily values are based on a 2,000 calorie diet. Your D.V. may be higher or lower depending on your calorie needs.

INGREDIENTS

2 c	all purpose flour
1¾ tsp	baking powder
¼ tsp	baking soda
1 can (14 oz.)	crushed pineapple
2¾ c	grated carrot
120 g	chopped walnut
65 g	shredded coconut
1 c	butter (unsalted)
1½ c	sugar
½ tsp	salt
4	large eggs
2 tsp	vanilla extract
1 tsp	ground cinnamon
⅜ c	milk (2%)

PREPARATION STEPS

1 In a large bowl, beat shaved butter for 1 minute on medium speed.

2 Add sugar and salt and cream on medium speed for 2 minutes.

3 Add eggs, vanilla extract, and cinnamon to mixture. Beat on medium speed for 1½ minutes.

4 Add milk and flour mixture to above mixture.

5 Blend with a spatula and then mix well for 2 minutes on medium low speed.

6 Fold in grated carrot, drained crushed pineapple, chopped walnut, and shredded coconut.

7 Blend on low speed for 1 minute.

8 Pour batter into prepared pan and bake for 55 minutes or until a skewer stick inserted into the centre of the cake comes out clean.

9 Cool cake on a wire rack for 15 minutes before removing cake from pan.

10 Enjoy!

CASSAVA CAKE

This is a less common cake. Once baked, the centre tends to be less brown than the outside. This gives the squares different coloured tones (i.e., brown, creamy, and hybrid), which appeals to different tasters. The various colours make cassava cake squares a most welcome bite-size party dessert.

- ■ Cassava
- □ Tougheners (butter & sugar, egg, salt)
- ■ Tenderizers (butter, coconut milk, sugar)

CHEMIST'S SECRETS

Refrigerate cassava cake for a couple of hours after cooling. This makes slicing into small pieces easier and reduces wetness. Also, it brings out a rather unique taste of the squares.

Cut into small bite-size square and serve in paper baking cups for parties.

SERVINGS PER RECIPE: 36
NUTRITIONAL INFORMATION PER SERVING

		% D.V.*
Calories	78.5	3.9
Total Fat	3.8 g	5.8
Sat. Fat	2.8 g	14.2
Cholesterol	11.0 mg	3.7
Total Carbs	10.9 g	3.6
Sugars	5.9 g	
Diet. Fibres	0.3 g	1.3
Protein	1.0 g	2.0
Sodium	19.2 mg	0.8
Calcium	18.0 mg	1.8
Potassium	65.8 mg	1.9

%D.V. = per cent daily values are based on a 2,000 calorie diet. Your D.V. may be higher or lower depending on your calorie needs.

PRE-PREPARATION

1 Defrost frozen grated cassava inside the package.

2 Grease and flour (or line with parchment paper) one 8"x 8"x 2" square pan.

3 Set oven to 350 °F.

INGREDIENTS

1 pkg.	frozen grated cassava (16 oz.)
¼ c	butter (unsalted)
½ c	sugar
⅛ tsp	salt
1	large egg
1 tsp	vanilla extract
200 ml	coconut milk
40g	shredded coconut
150 ml	sweetened condensed milk

PREPARATION STEPS

1 In a large-sized bowl, beat shaved butter for 20 seconds on medium speed.

2 Add sugar and salt and cream on medium speed for 1 minute.

3 Add egg and vanilla extract and beat for 1 minute on medium speed.

4 Add defrosted cassava, coconut milk, and shredded coconut to above mixture. Mix for 2 minutes on medium low speed.

5 Pour mixture into prepared baking pan and bake for 40 minutes.

6 Remove pan from oven and spread condensed milk evenly on top of cassava cake.

7 Bake for another 15 minutes or until the top has browned.

8 Leave cake in oven for another 2 minutes before taking it out.

9 Cut and serve in squares or rectangles.

10 Enjoy!

chocolate
CHIP CAKE

This is a rather simple cake to make. It is moist and has a great chocolate flavour. The presence of chocolate chips brings out a colour contrast, appealing to the eyes.

■ Flour
□ Tougheners (butter & sugar, egg, milk, salt, yogurt)
■ Tenderizers (baking powder/soda, butter, milk, sugar, yogurt)

CHEMIST'S SECRETS

🧪 Use sour cream if yogurt is unavailable.

🧪 Use Vanilla pudding mix for lighter cake colour

🧪 Coat semi-sweet chocolate chips with cocoa powder by shaking them in a zippered freezer/sandwich bag.

🧪 For larger contrast, use white chocolate chips instead.

PRE-PREPARATION

1 Sift and mix all purpose flour, chocolate pudding mix, cake flour, cocoa powder, baking powder, and baking soda in a medium-sized bowl and put aside.

2 Grease and coat with cocoa powder a 10" Bundt cake pan.

3 Set oven to 350 °F.

INGREDIENTS

1½ c	all purpose flour	⅛ tsp	salt
½ c	cake flour	5	large eggs
4 tsp	cocoa powder	1 tsp	almond extract
2 tsp	baking powder	1 tsp	vanilla extract
¼ tsp	baking soda	1½ c	yogurt (1.5%)
1 pkg	chocolate pudding mix	¾ c	milk (2%)
1 c	butter (unsalted)	150 g	small size semi-sweet chocolate chips
¾ c + 1 tsp	sugar		

SERVINGS PER RECIPE: 24
NUTRITIONAL INFORMATION PER SERVING

		% D.V.*
Calories	206.0	10.3
Total Fat	10.9 g	16.8
Sat. Fat	6.5 g	32.6
Cholesterol	66.0 mg	22.0
Total Carbs	24.6 g	8.2
Sugars	14.7 g	
Diet. Fibres	0.6 g	2.5
Protein	3.8 g	7.7
Sodium	141.8 mg	5.9
Calcium	70.4 mg	7.0
Potassium	71.7 mg	2.0

%D.V. = per cent daily values are based on a 2,000 calorie diet. Your D.V. may be higher or lower depending on your calorie needs.

PREPARATION STEPS

1 In a large glass bowl, beat shaved butter for 1 minute on medium speed.

2 Add sugar and salt and cream on medium speed for 2 minutes.

3 Add eggs, almond extract, and vanilla extract to butter mixture and beat for 1½ minutes on medium speed.

4 Add yogurt, milk, and flour mixture to above mixture.

5 Blend with a spatula and then mix well for 2 minutes on medium low speed.

6 Mix in chocolate chips for 1 minute on low speed.

7 Pour batter into prepared pan.

8 Bake for 60 minutes, or until a skewer stick inserted into the centre of cake comes out clean.

9 Let cake cool on a wire rack for 15 minutes before removing it from pan.

10 Enjoy!

lemon orange
CHIFFON CAKE

Chiffon cake is somewhere between an angel food cake and a sponge cake. The techniques for baking the other foam cakes apply equally here. It uses fewer eggs but more flour than the other cakes. The cake's texture is therefore firmer and it is more filling than the other cakes. The additions of fruit extracts and zests add wonderful aromas and a unique mouth feel.

■ Flour
☐ Tougheners (cream of tartar, egg white, salt, sugar)
■ Tenderizers (baking powder, egg yolk, milk, oil, sugar)

CHEMIST'S SECRETS

🧪 Separate egg white from egg yolk while egg is cold from the fridge.

🧪 Start preparation steps only when egg whites and egg yolks are at room temperature (about 30 minutes after separation).

🧪 Use superfine sugar for better results. Superfine sugar can be made from granulated sugar using a coffee bean grinder or a blender.

🧪 Avoid using plastic or wooden mixing bowls and spatulas as they absorb and retain fat, which is a "no no" for foam cakes.

PRE-PREPARATION

1 Separate eggs into whites and yolks and place them in two separate large-sized bowls.

2 Sift and mix flour and baking powder in a medium-sized bowl.

3 Grate lemon and orange zests and put aside.

4 Set oven to 350 °F.

INGREDIENTS

6	egg whites *(from large eggs)*	1 tsp	lemon extract
		1 tsp	orange extract
6	egg yolks *(from large eggs)*	1¼ c	sugar
		1¼ c	milk
2¼ c	cake flour	¾ c	vegetable oil
2¼ tsp	baking powder	¼ tsp	salt
1 tbsp	lemon zest	½ tsp	cream of tartar
1 tbsp	orange zest		

SERVINGS PER RECIPE: 20
NUTRITIONAL INFORMATION PER SERVING

		% D.V.*
Calories	191.1	9.6
Total Fat	10.2 g	15.6
Sat. Fat	1.2 g	6.0
Cholesterol	64.9 mg	21.6
Total Carbs	22.2 g	7.4
Sugars	13.4 g	
Diet. Fibres	0.1 g	0.3
Protein	3.2 g	6.5
Sodium	97.2 mg	4.1
Calcium	58.9 mg	5.9
Potassium	44.6 mg	1.3

%D.V. = per cent daily values are based on a 2,000 calorie diet. Your D.V. may be higher or lower depending on your calorie needs.

PREPARATION STEPS

1 Beat egg yolks, lemon extract, and orange extract on medium speed for 2 minutes.

2 Add sugar (1¼ c minus 1 tbsp) gradually and beat for 3 minutes on medium high speed.

3 Add and mix flour mixture and milk/oil alternatively to egg yolk mixture over 2 minutes on low speed, beginning with flour and ending with milk/oil.

4 Beat mixture for 3 minutes on medium high speed.

5 Add lemon zest and orange zest and mix on low speed for 1 more minute.

6 Add 1 tbsp of sugar and salt to egg whites. Beat for 1 minute on low speed and for another 2 minutes on medium low speed.

7 Add cream of tartar to egg whites and beat for 4 minutes on high speed.

8 Fold egg yolk mixture into beaten egg whites gently over 1 minute.

9 Pour final mixture into an un-greased 10" tube pan.

10 Bake for 50 minutes or until a skewer stick inserted into the centre of the cake comes out clean.

11 Invert and place cake pan on a wire rack (or on four glasses) to cool. Remove cake from pan after 1½ hours.

12 Enjoy!

mango COCONUT CAKE

For mango lovers, this cake is a must. The cake is rather straightforward to make. It is moist and has a strong taste of mango. Dried mango is preferred to fresh mango as it presents a unique sweet taste. Mango pulp adds mango colour to the cake and shredded coconut accentuates the cake's aroma.

■ Flour

□ Tougheners (butter & sugar, egg, salt)

■ Tenderizers (baking powder, butter, mango pulp, sugar, yogurt)

CHEMIST'S SECRETS

🧪 Use sour cream if yogurt is not available.

🧪 For those who desire a stronger mango flavour, use 125g to 150g (instead of 100g) of dried mango.

🧪 Similarly, use ½ c (instead of ¼ c) of mango pulp.

PRE-PREPARATION

1 Sift and mix cake flour, baking powder, and baking soda and set aside in a medium-sized bowl.

2 Cut dry mango into small pieces and put aside.

3 Weigh out 50 g of shredded coconut and put aside.

4 Grease and flour (or line with parchment paper) three (3) 3"x6"x2" loaf pans.

5 Set oven to 350°F.

SERVINGS PER RECIPE: 30
NUTRITIONAL INFORMATION PER SERVING

		% D.V.*
Calories	105.0	5.3
Total Fat	5.2 g	7.9
Sat. Fat	3.2 g	15.8
Cholesterol	29.5 mg	9.8
Total Carbs	14.5 g	4.8
Sugars	9.0 g	
Diet. Fibres	0.6 g	2.3
Protein	1.7 g	3.5
Sodium	61.6 mg	2.6
Calcium	28.6 mg	2.9
Potassium	19.7 mg	0.6

%D.V. = per cent daily values are based on a 2,000 calorie diet. Your D.V. may be higher or lower depending on your calorie needs.

INGREDIENTS

1 ¾ c	cake flour
1 ¾ tsp	baking powder
100 g	dried mango
50 g	shredded coconut
½ c	butter (unsalted)
¾ c + 1 tbsp	sugar
⅛ tsp	salt
3	large eggs
1 tsp	almond extract
¾ c	vanilla yogurt (1.5%)
¼ c	mango pulp

PREPARATION STEPS

1 In a large-sized bowl, beat shaved butter for 30 seconds on medium speed.

2 Add sugar and salt and cream on medium speed for 1 ½ minutes.

3 Add eggs and almond extract and beat for 1 minute on medium speed.

4 Add yogurt, mango pulp, and flour mixture to above mixture.

5 Blend with a spatula and then mix well for 2 minutes on medium low speed.

6 Fold in dry mango and shredded coconut and mix for 1 minute on low speed.

7 Pour batter into prepared pans and bake for 45 minutes or until a skewer stick inserted into the centre of the cake comes out clean. (Alternatively, use a 5"x9"x3" loaf pan and bake for 55 minutes.)

8 Cool cakes on a wire rack for 15 minutes before removing from pans.

9 Enjoy!

SPONGE CAKE

This is one of the most difficult cakes to make. However, if you could master this cake, you would have succeeded in baking one of the most delicious cakes in the world. The cake has an eye-pleasing creamy, lemon colour. It is very light and spongy which some describe it as a "melt in your mouth" feel and texture.

■ Flour

□ Tougheners (cream of tartar, egg whites, salt, sugar)

■ Tenderizers (baking powder, egg yolk, milk, oil, sugar)

CHEMIST'S SECRETS

Separate egg white from egg yolk while egg is cold from the fridge.

Start preparation steps only after egg whites and egg yolks are at room temperature (about 30 minutes after separation).

Use superfine sugar for better results. Superfine sugar can be made from granulated sugar using a coffee bean grinder or a blender.

Use different combinations of extracts such as almonds, lemon, and orange to experience a wide range of aroma.

Avoid using plastic or wooden mixing bowls and spatulas as they absorb and retain fat, which is a "no no" for foam cakes.

PRE-PREPARATION

1 Separate eggs into egg whites and egg yolks and place them in two separate large-sized bowls.

2 Sift and mix cake flour and baking powder and set aside in a small-sized bowl.

3 Set oven to 325 °F.

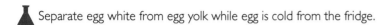

SERVINGS PER RECIPE: **20** NUTRITIONAL INFORMATION PER SERVING		
		% D.V.*
Calories	137.4	6.9
Total Fat	7.7 g	11.8
Sat. Fat	1.0 g	5.1
Cholesterol	85.3 mg	28.4
Total Carbs	14.4 g	4.8
Sugars	8.5 g	
Diet. Fibres	0.3 g	1.1
Protein	3.5 g	7.0
Sodium	67.8 mg	2.8
Calcium	30.9 mg	3.1
Potassium	32.3 mg	0.9

*%D.V. = per cent daily values are based on a 2,000 calorie diet. Your D.V. may be higher or lower depending on your calorie needs.

INGREDIENTS

1¼ c	cake flour
1¼ tsp	baking powder
8	egg whites *(from large eggs)*
8	egg yolks *(from large eggs)*
1½ tsp	vanilla extract
⅘ c	sugar
½ c	milk
½ c	vegetable oil
⅛ tsp	salt
½ tsp	cream of tartar

PREPARATION STEPS

1 Beat egg yolks and vanilla extract on medium speed for 2 minutes.

2 Add sugar (⅘ c minus 1 tbsp) gradually and beat for 4 minutes on medium high speed.

3 Add and mix flour mixture and milk/oil alternatively to egg yolk mixture over 2 minutes on low speed, beginning with flour and ending with milk/oil.

4 Beat above mixture for 4 minutes on medium high speed.

5 Add sugar (1 tbsp) and salt to egg whites. Beat for 1 minute on low speed and 2 minutes on medium low speed.

6 Add cream of tartar to egg whites and beat for 4 minutes on high speed.

7 Fold egg yolk mixture into beaten egg whites gently over 1 minute.

8 Pour final mixture into an un-greased 10" tube pan.

9 Bake for 40 minutes or until a skewer stick inserted into the centre of the cake comes out clean.

10 Invert and place cake pan on a wire rack (or on four glasses) to cool.

11 Remove cake from pan after 1½ hours.

12 Enjoy!

white chocolate STRAWBERRY CAKE

This cake builds on your success in making a sponge cake. Use an 8" springfoam pan instead of a 10" tube pan to bake in a taller cake. White chocolate flakes and fresh strawberries add contrasting colours to the cake, appealing to the eyes and mouth-watering taste to the taste buds.

■ Flour

□ Tougheners (cream of tartar, egg white, salt, sugar)

■ Tenderizers (baking powder, egg yolk, milk, oil, sugar)

CHEMIST'S SECRETS

⚗ To facilitate cutting sponge cake into two equal slabs, insert four tooth picks on the side mid-way from the top and bottom to serve as guides.

⚗ Use a sharp knife (such as an electric knife or a bread knife) to cut across cake guided by the tooth picks' positions.

⚗ Dip and coat eight (8) whole strawberries with melted white chocolate (70 g) by using a double boiler to melt chocolate.

⚗ To increase contrast, dip whole strawberries in melted dark chocolate instead.

PRE-PREPARATION *

1 Using the sponge cake recipe in this section, prepare a cake with an 8" springfoam cake pan.

2 Slice the cake side way to result in two equal round slabs.

3 Wash and dry 15 strawberries.

4 Slice seven (7) strawberries and keep eight (8) whole strawberries for later use.

5 Grate/shave 150 g white chocolate with a grater (or a food processor) and put in a small-sized bowl.

6 Beat heavy cream (together with ½ a pouch of gelatin, ¼ tsp of vanilla extract, and 1 tbsp of sugar) on high speed in a medium-sized bowl (until stiff peaks form).

7 Keep cream in refrigerator for later use.

* In collaboration with Derek Christopher

SERVINGS PER RECIPE: 16
NUTRITIONAL INFORMATION PER SERVING

		% D.V.*
Calories	335.9	16.8
Total Fat	23.5 g	36.2
Sat. Fat	10.0 g	50.0
Cholesterol	148.9 mg	49.6
Total Carbs	27.0 g	9.0
Sugars	18.0 g	
Diet. Fibres	0.9 g	3.7
Protein	6.2 g	12.4
Sodium	104.0 mg	4.3
Calcium	80.2 mg	8.0
Potassium	131.4 mg	3.8

*%D.V. = per cent daily values are based on a 2,000 calorie diet. Your D.V. may be higher or lower depending on your calorie needs.

INGREDIENTS

1	sponge cake *(made in a 8" springfoam pan)*
16 oz.	large, fresh strawberries
150 g	white choclate
2 c	heavey cream
1 pouch	gelatin
¼ tsp	vanilla extract
1 tbsp	sugar

PREPARATION STEPS *

1 Lay first slab of sponge cake on a flat surface.

2 Spread ½" layer of cream on slab.

3 Put sliced strawberries on cream layer.

4 Cover sliced strawberries with cream.

5 Place second slab of sponge cake on top of cream.

6 Spread cream on cake top and side.

7 Coat side of cake with grated chocolate.

8 Begin decorating cake overlapping sliced strawberries working from outside in.

9 Finish decorating by placing eight (8) whole strawberries in the centre.

10 Glaze whole and sliced strawberries with gelatin (½ pouch in 2 c of water).

11 Refrigerate for two hours before serving.

12 Enjoy!

   * In collaboration with Derek Christopher

ABOUT THE AUTHOR

Dr. Walter H. Chan is a physical chemist by training. Walter picked up cake baking as a therapeutic hobby in year 2000 when he was in fulltime study while holding a fulltime job. He quickly recognized many issues with baking and he was determined to learn about its basics and address these challenges.

Walter obtained B.Sc. (Honours) and M.Sc. degrees from Queen's University (Canada) and was awarded a Ph.D. degree from the University of Western Ontario (Canada). He conducted post-doctoral research at the Ohio State University (USA). He also holds a Master of Theological Studies degree from Tyndale Seminary in Toronto (Canada).

Walter was a former director with the Ontario government and an Adjunct Professor at the University of Toronto. Walter is also the author of a recent book titled, "Choosing Work-Life Balance: The Keys to Achieving What Many Think is Unattainable" (Xlibris 2010; http://worklifebalance.webstarts.com).

CPSIA information can be obtained
at www.ICGtesting.com
Printed in the USA
255758LV00002B

9780986722004